LIVING PHILOSOPHY

Educated in Warsaw and at New College, Oxford, where he received a D.Phil. in 1964, Henryk Skolimowski has been actively engaged in healing the planet for the last fifteen years. He is the creator of eco-philosophy and the director of the Eco-Philosophy Center. His book *Eco-philosophy: Designing New Tactics for Living* was the first in the field; it has been translated into twelve languages. He runs workshops on eco-philosophy and eco-yoga during the summer at the village of Theologos, on the island of Thassos in northern Greece. He has published twelve books and over 200 articles. He is presently teaching at the University of Michigan, Ann Arbor, and all over the world. In the autumn of 1991 he was appointed Head of the Chair of Ecological Philosophy (the first in the world) at the Technical University of Łódź, in central Poland. His main interests are eco-philosophy, eco-ethics and evolutionary epistemology. He is presently working on the participatory theory of mind and on eco-religion.

LIVING PHILOSOPHY

ECO-PHILOSOPHY AS A TREE OF LIFE

HENRYK SKOLIMOWSKI

ARKANA

ARKANA

Published by the Penguin Group
27 Wrights Lane, London W8 5TZ, England
Penguin Books USA Inc., 375 Hudson Street, New York, New York 10014, USA
Penguin Books Australia Ltd, Ringwood, Victoria, Australia
Penguin Books Canada Ltd, 10 Alcorn Avenue, Toronto, Ontario, Canada M4V 3B2
Penguin Books (NZ) Ltd, 182–190 Wairau Road, Auckland 10, New Zealand

Penguin Books Ltd, Registered Offices: Harmondsworth, Middlesex, England.

Published by Arkana 1992
1 3 5 7 9 10 8 6 4 2

Printed in England by Clays Ltd, St Ives plc
Set in 10/12½ Monophoto Baskerville

Contents

Preface

Among twentieth-century thinkers, Albert Schweitzer was most aware of the importance of the right world-view (*Weltanschauung*) for a civilization to survive and prosper, as he repeatedly wrote that 'We should all recognize fully that our present entire lack of any world-view is the ultimate source of all catastrophes and miseries of our times.' Yet Schweitzer did not provide a new world-view, except one element of it – reverence for life. In contrast, the present book offers a completely new world-view, from cosmology to consciousness, as a foundation for a restoration of our civilization. Ecological philosophy, here outlined, spells out not only new tactics for living but a new comprehensive ecological world-view on which these new tactics are based.

Introduction

After all the juggling with economic figures is done, there is still life to be lived. The meaning of life is not to be derived from any economic calculation: its roots lie far beyond all economic and physical parameters. Bertrand Russell and other positivists of the twentieth century have nearly persuaded us that the human project is to explore the physical world. Eco-philosophy insists that the human project is a rediscovery of human meaning, related to the meaning of the universe.

In exploring the physical world we have created complicated matrices on to which we map the welter of physical phenomena. The detail of these matrices has become so subtle, complex and exhaustive that we simply have no room for understanding other things, such as human meaning.

One aim of this book is to unravel the variety of mechanistic and physical relationships (within which we are wrapped and by which we are defined) in order to demonstrate that most of our crises, particularly the economic ones, do not arise as the result of mismanagement, ill will or the insufficiency of rationality in our approaches; they arise for more fundamental reasons, because *we have constructed a deficient code for reading nature, leading to a deficiency in interacting with nature.* The root cause lies in the very foundations of our scientific world-view and in the very perceptions that this world-view engenders.

Alternative lifestyles do not only require living differently but also *knowing* differently. We must be able to provide a rational justification for our new lifestyles, which will amount to nothing less than providing a new rationality. We must be convinced in our hearts and minds that frugality is not a depressing abnegation and self-denial but an act of positive manifestation of new qualities; only then will it become *elegant frugality*. Alternative life-styles must therefore signify not only changes in our technology, economics and patterns of living, but changes in our morality, rationality and conceptual thinking.

On the philosophical level this book provides an outline of a new metaphysics, which consists of a new reading of the cosmos, of evolution, of human nature – all woven into one coherent framework. Yet it must be emphasized that the overall aim is to provide a new purpose, a new inspiration and a new hope for humankind. In this sense, eco-philosophy is a philosophy of life.

After having witnessed the slow death of positivist philosophy over the last decades, and of Marxist philosophy more recently, the time has come for releasing our philosophical imagination once more in order to provide the foundation of a new intellectual understanding of the world around us, which has been in tatters for decades.

Eco-philosophy is a rational restatement of the unitary view of the world in which the cosmos and the human race belong to the same structure. In evolving eco-philosophy, we are not throwing out the best aspects of the Western rational tradition. We are, however, transcending the crippling effects of mechanistic rationality. *No new world-view can be irrational or anti-rational.* What we need to evolve is an alternative rationality. This is what, among other things, eco-philosophy endeavours to offer: a form of rationality that does not offend reason but, on the contrary, celebrates and eulogizes it in much more magnificent a manner than the mechanistic paradigm would ever allow.

Another aspect of eco-philosophy is *reverential thinking*. As we interact with the world differently, we must be able to think about it differently – by beholding the earth and all its creatures in a reverential and compassionate way. Thus we need to transform our present mechanistic consciousness so that it becomes *ecological consciousness*. Reverential thinking and reverential perception must pervade our system of education, our institutions and our daily lives. Only then will ecological consciousness become a reality.

The twenty-first century will be an ecological century or we may not have the twenty-second century at all. Some people have suggested that 'Environmentalism will be the next major political issue, just as conservatism and liberalism have been in the past.' This is a large claim; but perhaps, at the same time, not large enough. For this claim still reduces ecology to ideology. Ecology is nowadays more than an ideology; conceived as an overall ecological perspective on all issues of life, it now assumes the role of a new religion, with ecological thinking emerging as quasi-religious thinking. We need a form of religious fervour to unite our energies and wills in the stupendous undertaking of saving the earth and ourselves in the process.

Ecology is the single unifying thread that ties together the whole planet and all its people. Eco-philosophy is an expression, in philosophical terms, of the new unity between humans, the planet and all other beings.

Some readers may worry about *how* to implement the programme of eco-philosophy. My answer is: let us get our thinking straight, our visions clear and our values relevant, to the ends of our lives. Then we shall find the ways and means of building a sustainable and radiant future. Thinking well, thinking with a high purpose and within the right understanding of the universe, is very important in our times. Logos is a very subtle and all pervading form of praxis.

Eco-philosophy, presented in this book, is like a tree. Out of the roots of eco-cosmology grows the trunk and branches of eco-philosophy – all organically linked. The tree is crowned with ecological consciousness, which in a subtle way feeds back into the roots. Thus the cycle is complete and self-renewing within itself.

I

Eco-cosmology as a New Point of Departure

ONE COSMOLOGY OR MANY COSMOLOGIES?

Cosmology is an ancient subject. The dawn of philosophy was in fact the dawn of cosmology. When Thales, Anaximander, Empedocles and other Greek philosophers in the sixth and fifth centuries BC began to abandon mythology and started to think in a new way, this led to a new understanding of the universe. And this was the beginning of both philosophy and cosmology.

Early Greek cosmologies were speculative and rather naïve. Let us be quite clear that any cosmology that tries to understand the structure and the origins of the universe is bound to be speculative. And so is present physical cosmology, which has been vigorously developed by astrophysicists during the last thirty years.

Stephen Hawking asks in his book, *A Brief History of Time*, a number of tantalizing questions such as: where did the universe come from and where is it going? did the universe have a beginning, and if so what happened *before* that? Such questions cannot be answered scientifically for they are metaphysical in nature. The assumptions on which science is based belong to metaphysics and not to science. Hence, the basic cosmological questions that present astrophysics explores *are* metaphysical and not scientific.

Yet many try to maintain that cosmology as elaborated by astrophysicists is legitimate science, while older cosmologies were speculation. Thus John Barrow writes:

> Cosmology is the science of the universe – its size, age, shape, wrinkles, origin, and contents. Mankind's oldest

> speculation, it has been transported in the twentieth century from the realm of metaphysics into the domain of physics, where speculation is not unbridled and where ideas must confront observations.[1]

Scientists have a tendency to call any product of their thinking 'scientific', even if the subject is clearly metaphysical, such as the origin of the universe. It is metaphysics *par excellence* according to the straight Aristotelian definition: *ta meta physica*, that which is beyond physics.

Barrow suggests that 'ideas [of present cosmologists] must confront observations'. This is a gratuitous claim. Past cosmologies always had to 'confront observations' – even the most primitive of cosmologies. If Barrow wishes to insist that physical cosmology has to confront *different* kinds of observation, then I shall agree.

Let me underscore a main point. In their very nature, past and present cosmologies are similar in structure. They are speculative and highly conjectural. Given the inscrutability of the phenomena they wish to explain (the origins of the universe), these speculations are metaphysical.

Is nothing therefore special about present physical cosmology? If Barrow and others wish to insist that there is – namely, that physical cosmology gives us the truth about the universe – then they are begging the question. For they are *assuming* that their cosmology is right, while other cosmologies are not. This is not a scientific claim, it is a metaphysical one. If we wish to remain strictly scientific, then we cannot make such a claim.

How to choose a metaphysic is a thorny question. One is inclined to agree with Willis Harman, who writes:

> It is futile to seek through research to answer the question 'What metaphysic is correct?' The basic reason is that *the research methodology itself grows out of a metaphysics*, so the research tends to lead us the full circle, back to that metaphysics.[2]

Nobody has a monopoly on the term 'cosmology'. There are

many different cosmologies as there are many different philosophies. Empirical philosophy, or physicalism, is one kind of philosophy. Platonism is another kind of philosophy. It would be absurd to deny the name of philosophy to Platonism because some philosophers, inspired by science, wish to consider physicalism as the only genuine philosophy.

Even within the realm of the present science there are quite different cosmologies, or at least rudiments of different cosmologies. David Bohm is a physicist known for his work in quantum theory, but also for his speculative theories about the nature of the universe. He contends that we live in an unfolding universe. It unfolds in a very special way – by making the transition from the *implicate* order to the *explicate* order. The implicate order is the state of things *in potentia*, and this order becomes the explicate order as events unfold.

What is of real importance is not the postulated transition from potence to actuality – many philosophers have proposed similar ideas – but the way things are envisaged in the implicate order. According to Bohm, all things are connected in a most fundamental, primordial, cosmological sense. ('The entire universe is basically a single indivisible unit.') Hence, elementary particles (in the double split experiment, for instance, in which changing the spin of one particle simultaneously changes the spin of a paired particle, whatever the distance between the two particles) can somehow 'know' what other particles are doing. The notion of an implicate order, when sufficiently spelled out, becomes a new cosmology. (More on Bohm pp. 20–22).

The order of the universe has fascinated philosophers and astronomers for millennia. While constructing various cosmologies, or attempting to understand at least part of the architecture of the cosmos, the philosophers and astronomers of the past were as partial to the idea of truth as present astrophysicists are. But they were also partial to the idea of beauty.

Among the great astronomer-philosophers who speculated on the nature of the order of heaven was, of course, Copernicus. For him the order of the universe was both perfect and beautiful – it could not be otherwise, as the universe was created by a perfect God. Copernicus wrote in *De Revolutionibus*:

> And what is more beautiful than the heavens which contain everything that is beautiful? The names themselves *Caelum* and *Mundus* are an evidence, of which one signifies purity and ornament and the other a work of sculpture. It is because of its exceptional beauty that many philosophers called the heavens simply the visible deity.[3]

It was not unusual for the great scientists and philosophers of the past to see beauty in the architecture of the cosmos and to invest the cosmos (and hence cosmologies describing the universe) with purpose, meaning and intention. Such is not the case with the present physical cosmology – but this cosmology is just one of many!

Thus, let us be perfectly aware that twentieth-century astrophysicists neither discovered nor invented the idea of cosmology. They have simply appropriated it for their own usage. Often they seem to give the impression that cosmology did not exist before they invented it, despite the fact that it is a noble and ancient discipline. True enough, when Socrates and Plato came on to the scene, a fundamental shift occurred in philosophy – cosmology ceased to be the centre of philosophical concern and, instead, the philosophy of man became the focus. Yet cosmology did not disappear either from Western philosophy or from the systems of thought of other cultures. Any coherent or even semi-coherent system of beliefs, which attempts to explain the structure and the origin of the universe and our relationship to the universe, is a cosmology. We can therefore legitimately and justifiably talk about the cosmology of the Hopi Indians or the cosmology of the Amazon Indians; and indeed, perceptive studies have been written on these cosmologies.[4]

In the late nineteenth century and the first decade of the twentieth century, the term cosmology was hardly used in physics and astrophysics – indeed, the latter barely existed at the time. Yet anthropologists used the term extensively and meaningfully to explain the belief systems of other people and other cultures.

To reiterate the point, we must not presume that the present day, scientific explanation of the structure of the universe is the only legitimate explanation, and that scientific cosmology is the only legitimate cosmology. As I have said, any semi-coherent system of beliefs that tries to explain the structure of the universe, and to explain our place within it, is a cosmology. Historically, therefore, we have had, and still have, hundreds of different cosmologies; each is legitimate in its own right.

In the twentieth century, before astrophysicists rallied round the term, 'cosmology' had been resuscitated in philosophical literature by Teilhard de Chardin, particularly in his opus *The Phenomenon of Man* (1957). Evolution is, in Teilhard's thinking, the focal point for understanding the origins, the structure and the meaning of the universe. Evolution is that process which, with a consummate skill, creates new options, new forms of life through which the transformation of matter into spirit occurs. Eco-cosmology, proposed in this chapter, builds on Teilhard's cosmology and yet goes beyond it – as evolution always does.

WHY DO WE NEED A NEW COSMOLOGY?

We need a new cosmology for a more fundamental reason than we are usually aware of. We need it as a new matrix for our action. We need it because our action, performed in the present framework, *continuously misfires*. In this section I will relate action to cosmology and will show that cosmology (this abstract underpinning of our thought), is very concretely linked with action, via values and philosophy.

Action is never mere action as such; it must, of necessity, be informed and guided. To act does not mean simply 'doing'. More importantly still, acting does not signify that form of activity which results in self-destruction. Thus, by action I am referring to purposeful and meaningful action. Purposefulness and meaningfulness are the attributes of action contained in its very conception. Now, what is purposeful and meaningful is not determined by action itself. Action is the *executor* of the goals and purposes conceived prior to action. Thus action is directed and guided by goals and desiderata that originate in the sphere outside action itself. What is the nature of these goals and desiderata? This question I want to examine in some detail as it is important for understanding our future and also understanding our present.

The various calls to action, and the various ways of chastising reflection as idle doing, make sense only if we know what action is about. Yet *we can only know what action is about through reflection.* Unreflected action is mindless doing, is thrashing around, or worse, maybe a destructive action. These simple truths must be reiterated in our age which is dominated by pragmatic philosophies within which action is the king and reflection a pauper. These pragmatic philosophies are based on insufficient reflection. Present technology, present applied science and present economics are equally based on insufficient reflection. Nobody questions the good intentions of technology. Nobody questions its positive achievements. But we would be unreflective fools if we did not see the negative consequences of technology, the poisonous debris that is left behind the chariot of technological progress. Technology has trivialized our lives; it has robbed our lives of quality and replaced spirituality with cheap gadgets.

The point is that many actions have been conceived in a limited frame of reference. Within this limited frame, they appear to be meaningful and purposeful. Only when we examine their consequences in larger frames of reference, and over longer stretches of time, do they appear counter-productive.

This situation may seem confusing to the practical individual, who is just a doer and does not engage in larger reflection. We may feel compassion for such people, but nevertheless hold them accountable for the consequences of their activities in the long run. Such is the meaning of responsibility: we are held responsible for the delayed consequences of our action, as well as for the immediate consequences. If you give arsenic to another person in small doses, although none of the doses kills by itself, by the accumulative effect over a period of time you will kill nevertheless. And you will be charged with murder.

Our practical individual symbolizes the whole technological civilization. This civilization is very impatient in examining the long-term consequences of its activities. Yet, this very civilization seems to be administering to itself small doses of arsenic daily.

At this point in history, we need to go beyond the worn-out cliché of our times, which calls people of action heroes and which considers philosophers bums. In going beyond the present cliché, we need to go beyond the *consciousness* that creates the cliché. We need to transcend the narrow pragmatism that so often has generated unreflected action. We need to re-examine our values, for values inform and guide our action.

In the broadest sense, the values that have guided meaningful action in human societies through the millennia are the ones that aim at increasing human happiness or decreasing human misery; at increasing justice or decreasing injustice; at making our lives more beautiful or decreasing ugliness in our lives; at increasing our knowledge and enlightenment or decreasing our arrogance and prejudice; at bringing heaven to earth or diminishing hell on earth. Thus truth, goodness, beauty, enlightenment and a sense of grace are the values that have always motivated meaningful action.

Technological society added to these values some new ones: *efficiency*, *control* and also *power*. These new values often seem to be in conflict with the old values.

Obviously what will be deemed as meaningful action will be different if, on the one hand, the values of beauty and harmony guide our actions, and, on the other hand, they are guided by the values of efficiency and power.

For Faustian Man, who believes that he only lives once and therefore is entitled to everything he wants (at whosoever's and whatever expense), the exploitation, even the plunder of the natural environment is a *convenience* required by his high living. For ecological individuals, who understand the interconnectedness of all things and the frail balances that prevail, and who feel responsible for future generations, this 'convenience' is a crime. The ecologist attempts to cultivate frugality; the Faustian is at the mercy of indulgence. And these two attitudes, frugality and indulgence, spell out different philosophies; each represents a different conception of what life is and should be about.

We have moved from the meaning of a particular action to the meaning of action as such; then to the values underlying various types of action and activities; then to philosophy underlying and engendering these values. These philosophies merge with cosmologies; often they are particular articulations of various cosmologies.

The main purpose of the discussion in this section is to establish that there is a link between cosmology and action. This link is mediated by two intermediaries, values and philosophy, but it is there none the less.

Cosmology ⇄ Philosophy ⇄ Values ⇄ Action

As we read the universe so we act in it. If we read the universe incorrectly, we will act incorrectly. How do we know that we have acted incorrectly? By the life that will result from the residue of our actions; the final test of our cosmology is what kind of life it engenders.

We are now returning to the main question of the section: why do we need a new cosmology? Because our action misfires; misfires on the level of the whole culture, on the level of the whole globe. It is not readily understood, if at all, that to mend such action we need something other than action of a very similar kind. For these two very similar kinds of action are usually guided and directed by very similar values and visions. If action continually misfires, we need to go deeper into the underlying matrix in order to realize that inappropriate values may be the reason. These values stem from certain philosophies, which in turn are influenced and determined by certain conceptions of the cosmos – which are, of course, our cosmologies.

Thus the manner in which we interpret the cosmos – what is it? what does it contain? how has it come about? what is its destiny? what is our place in it? – filters down to the level and meaning of our actions. (As we read the universe so we act in it.)

Cosmology is the last link of the chain. Cosmology justifies ultimately all other endeavours and itself needs no further justification, except retroactively – by the consequences it brings to our lives. For some people, of course, religion is the last link of the justification chain, but religion is simply another form of cosmology.

The choice of our cosmology determines for us not only the image of the world but also the meaningfulness of our actions. For cosmology defines not only the physical universe outside ourselves, it also indirectly defines our place in it. If we assume that the universe is nothing but physical matter, we have great difficulty accommodating spirituality within it. If the universe is assumed to be divine, our spirituality follows naturally – as an inherent aspect of this universe, not an anomaly. If the universe is assumed to be orderly and harmonious, we are encouraged and invited to envisage our lives as harmonious and connected. If the universe is assumed to be chaotic, or worse, a garbage pit, we are allowed, and in a sense encouraged, to look at our lives as worthless; or worse still – as garbage reflecting the garbage of the universe. (This, unfortunately, may be the case with many lives, although people are not quite aware of how the low image of themselves is a consequence of the poor image of the universe imposed on them.)

Our lives are the mirrors in which the fundamental characteristics of the universe, as we understand it, are reflected.

Thus, if we wish to insist on the meaningfulness of our lives, on their purpose and beauty, we had better assume that the universe has meaning, purpose and beauty. Even if we cannot *prove* it with regard to the physical universe, it helps us to maintain the coherence and meaning of our own lives. *To assume an inherent purpose of the physical universe is thus a methodological imperative that helps us to govern ourselves in the human world.*

It may be said that I cannot prove my assumption. In response I can say that nobody can *disprove* it in the strict scientific sense. In any case, this is the assumption that the traditional cosmologies have made, namely that the universe is purposeful, meaningful, beautiful, hospitable and sympathetic to our striving. This is what eco-cosmology also assumes, namely that the universe is home for the human race, and we are its stewards, custodians and guardians. Moreover, given our unique role in the universe, given the creative nature of the mind, it is safe to assume that we co-create with the universe and contribute to its destiny. Before we go into details of eco-cosmology, let us take a brief look at the legacy of the mechanistic cosmology.

THE LEGACY OF THE MECHANISTIC COSMOLOGY

The metaphysical and cultural reconstruction of our time is not merely tinkering with environmental problems (important as they are), but addressing ourselves to the fundamental causes underlying our multiple crises. These causes go beyond the economic and technological. They even go beyond the moral. These crises are embedded in the underlying matrix of our world-view, our cosmology.

What has been backfiring on us are the shortcomings of our cosmology, of our world-view, which is now functioning as a strait-jacket. *The mechanistic cosmology at present provides a deficient code for reading nature. Hence our deficiency in interacting with nature.* The mechanistic cosmology, with its abstract non-compassionate rationality, provides an inadequate basis for social and human orders. Hence, the variety of rational models, evolved under the auspices of scientific rationality, so often are part of the problem, not a solution to human and social dilemmas.

Our world-view and our lifestyles are intimately connected. The mechanistic conception of the universe, in the long run, *implies* and *necessitates* a human universe that is cold, objective and uncaring. As a consequence, human meaning atrophies.

This point needs to be quite clearly spelled out, namely, that the atrophy of meaning and the triumph of the quantity are closely related. To put it another way, human meaning and cold number do not cohabit well together. The very language of science and its categories do not allow for the expression of the meanings of our humanness.

Thus, the atrophy of human meaning in the mechanistic system is not the result of benign neglect. Rather it is an *essential conse-*

quence of the mechanistic cosmology. The philosophical codification of the mechanistic cosmology is the doctrine called empiricism. David Hume in his *Enquiry Concerning Human Understanding* is perhaps the best expositor of this doctrine. Hume wrote:

> When we run over libraries, persuaded of these principles [of empiricism], what havoc must we make? If we take in our hand any volume; of divinity or school metaphysics, for instance; let us ask, *Does it contain any abstract reasoning concerning quantity or number?* No. *Does it contain any experimental reasoning concerning matter of fact and existence?* No. Commit it then to the flames: for it can contain nothing but sophistry and illusion.[5]

This is a classic paragraph. The philosophy of empiricism put in a nutshell. And devastating to theology, to metaphysics and actually to all philosophy! Also to the meaning of human life, as well as to Hume's treatise itself. If we were to take Hume at his word, we would want to throw his *Enquiry* into the flames – as it is concerned with neither quantity nor number, but with metaphysical speculations about each.

The ghost of Hume has been haunting all the edifices of so-called rational knowledge. A curious paradox is that we want to be good empiricists, regardless of whether we understand the consequences of empiricism or not. We all want to base our discourse and reasoning on fact and number, because such are the dogmas of our present cosmology.

The influence of the mechanistic cosmology is still paramount in the present Western society. We know that this cosmology is inadequate. We know that some of its consequences are pernicious. We know that an unmitigated pursuit of objectivity is a somewhat paranoic quest. We also know that the alienation, atomization and decimation of society, of natural habitats, of individual human existences, are partly the result of a structure of knowledge that incessantly atomizes, isolates and separates.

We have made various attempts to ameliorate the situation. However, the main imperatives of our cosmology are still holding us in their grip: to quantify, to objectify, to 'thingify'. What is therefore of great importance is not only a thorough examination of the nature of this cosmology, but an imaginative endeavour to create and spell out alternative cosmologies that would provide an antidote to the mechanistic cosmology, an alternative vision of the universe, as well as a set of alternative strategies for its exploration, including alternative modes of thinking and alternative modes of justification.

Eco-cosmology attempts to be this kind of endeavour. It not only wishes to critique the existing cosmologies, but it wishes to construct a new cosmological scaffolding, a new matrix, through which we can interact with the cosmos and ourselves in a new way; and within which 'quantity' and 'number' and 'experimental reasoning concerning matter of fact' are to be confined to their proper place, a rather modest place, and not worshipped as deities.

Our civilization is at a new juncture, at a crossroads, and we need a new cosmology so that we can get somewhere. To reiterate, cosmology provides the roots out of which a multitude of things grow. To paraphrase T. S. Eliot, 'A wrong conception of the universe implies somewhere a wrong conception of life, and the result is the inevitable doom.'

THE STRUCTURE OF ECO-COSMOLOGY

The architecture of eco-cosmology is supported by seven main pillars:

1 The anthropic principle

Eco-cosmology accepts the fact of the existence of the physical universe, which emerged from its primordial, mysterious beginnings some 15 billion years ago. Eco-cosmology accepts the essential mystery of the very origin of the universe. This mystery is part of the beauty of the universe.

Eco-cosmology accepts the conclusions of present astrophysics concerning the size, density and properties of the physical universe in its cosmic evolution. These conclusions lead us to believe that the phenomenon of life is not only possible but perhaps inevitable – that is if we closely examine so-called cosmological constants which are responsible for the structure of the universe and its unique patterns. Freeman Dyson writes:

> As we look out into the universe and identify the many accidents of physics and astronomy that have worked together to our benefit, it almost seems as if the universe must in some sense have known that we were coming.[6]

During the last twenty years we have learnt to look at the universe in a new way, and to ask some revealing questions. Why is the universe as it is? Because we are here. It is indeed a staggering realization that the composition of the universe is so exquisitely balanced that it makes life not only possible, but perhaps necessary. This insight has led to the formulation of the anthropic principle.[7] The anthropic principle, simply stated, maintains that the fate of the universe is bound with the fate of the human (*anthropos*). This principle has many formulations, one of which is that the universe is so constructed that it was bound to bring about intelligent life.

As we search deeper and deeper into the underlying structure of the cosmic evolution, we are more and more convinced that the 'coincidences' may not have been so coincidental but rather fragments of a larger pattern. As our knowledge and hypotheses grow subtler and deeper, the universe seems to be revealing to us its subtler and deeper features. This is how it ought to be: the subtler the mind, the subtler are the phenomena it discovers.

The anthropic principle, translated into the language of eco-philosophy, means that the universe is home for the human. We are its legitimate dwellers, not some kind of cosmic freaks. In a sense we are its justification. The awesome cosmic changes can be explained simply by the necessity of bringing about life endowed with intelligence. Yet, on another level, the conception of the universe as home for humankind – with us as its custodians – implies that we are responsible for our fate and for all there is.

2 Evolution conceived as the process of creative becoming

This is the view of evolution that Teilhard de Chardin holds – evolution as an ever-growing process of complexity, in the wake of which new layers of consciousness emerge. The complexity-consciousness thesis, explaining the main modus of evolution, does not necessarily entail the idea of predetermined design; or the idea of God, who is brought through the back door. But the thesis does challenge the claims of narrow Darwinians who surmise that evolution is a stupid monkey, who sits at a typewriter and by pure chance (given almost infinite time), types all the works of Shakespeare. No, evolution is more subtle than that. If the anthropic principle may be said to be a force endowed with some intelligence, then evolution is no less intelligent.

Let us make an important connection. Evolution as a process of incessant becoming can be seen as a more precise articulation of the anthropic principle. Teilhard was not aware of it – nor could he be, as he wrote his opus in the 1950s. The chief proponents of the anthropic principle do not seem to be aware of it either. But here it is: after the cosmological constants, as understood by the anthropic principle, had done their work (during the first twelve billion years of the cosmological evolution), the next stage of this evolution required a new vehicle; a specific vehicle to articulate life out of the well-established cosmological/chemical niches. This vehicle is evolution. *Conceiving of evolution as a creative force, which ceaselessly articulates life in ever new forms of consciousness, is not only congruent with the anthropic principle, but a necessary extension of it.* Thus evolution continues the work of the anthropic principle!

Teilhard's reconstruction of evolution is compelling enough.[8] But it becomes even more compelling when we realize that evolution is an aspect of and the continuation of the work of the anthropic principle. In order not to come to a grinding halt, the anthropic principle had to conceive of a force to articulate life *more explicitly*. This force is evolution.

Thus the anthropic principle articulates itself through creative evolution. Creative evolution represents a continuation of the early works of the forces of the cosmological constants. We have now connected the first twelve billion years of evolution (the anthropic principle) with its next four billion years (creative evolution). It is one unfolding process.

3 The participatory mind

A cosmology that only gives us a picture of the world, excluding the human being, and which does not explain how we interact with what is out there, is essentially incomplete. Eco-cosmology, while accepting the tenets of the anthropic principle, and of the heritage of creative evolution, also outlines a theory of the participatory mind. The participatory mind lies hidden in the layers of unfolding evolution. The immediate predecessor of the participatory mind is John Archibald Wheeler's conception of the participatory universe. Wheeler writes:

> The universe does not exist 'out there' independent of us. We are inescapably involved in bringing about that which appears to be happening. We are not only observers. We are participators. In some strange sense this is a participatory universe.[9]

The idea of the participatory mind occurred to me while I was simultaneously contemplating two trends of thought: the legacy of Teilhard's vision of evolution, in which the role of the mind is somewhat neglected; and the legacy of the consequences of present astrophysics, in which the universe is conceived as participatory, and yet the mind is not present. I realized that the participatory mind is the element which makes sense of the participatory universe and which is indispensable for the understanding of evolution as the process of becoming through the increase of consciousness.

The conception of the participatory mind maintains that mind is present in *all* products of our knowledge and in *all* pictures of the world. We are bound by the *noetic condition* – by the presence of our mind in all the forms of our knowledge and of our understanding. Whatever we receive from the world is filtered through our mind. If it is not filtered, it is not received. If we were a different species, and possessed an altogether different structure of mind, our picture of the world and all our understanding of it would be different. Therefore, while bound by our mind, which is the shaper of reality, we can never describe the cosmos as it is. *We always partake in what we describe.* Our description is a fusion of our mind and 'what is there'. Our mind invariably and tirelessly elicits (through its various faculties and sensitivities) from the amorphous primordial data of the universe.[10]

When evolution became conscious of itself it heralded the arrival of self-consciousness. As self-consciousness articulates itself, it begins to perceive that it is participatory consciousness, which co-creates with the universe and completes the meaning of the participatory universe. The anthropic principle thus was bound to articulate itself in the form of participatory consciousness that we call the participatory mind.

This point deserves emphasis: data is never data as such, it is always mediated, influenced, moulded, shaped and determined by the mind. This is the meaning of the participatory mind. If anything is registered, let alone formulated and articulated in our mind, or better still, expressed in the language and the annals of our knowledge, it is already filtered and structured through the mind.

The idea of the participatory mind not only frees us to see the universe in a new way; it also enhances the freedom and dignity of the human individual. To give justice to our participatory mind compels us to co-create with the universe. The awareness of the creative power of our mind only emphasizes our responsibility for our own lives, for the fate of the earth, for the fate of the universe. The universe is responsible for our birth, we are now responsible for its fate.

4 The implicate order

Another part of the overall structure of eco-cosmology is what

David Bohm calls the implicate order, a principle similar to the anthropic principle. It attempts to convey some of the essential characteristics of the universe in its unfolding state. Our language is invariably a language of parts, very good to describe atoms, but not genuinely adequate to describe complex wholes, let alone the universe in its evolution. Therefore David Bohm and his followers use analogies. One of the analogies that Bohm and others use to convey the meaning of the implicate order is the following: suppose we drop a spot of ink on the top of the smooth surface of a cylinder of glycerine. We now rotate the cylinder around its axis. The drop becomes a smudge, then disappears – at least to our eyes. But it is there. If we rotate the cylinder back, we will bring it back to existence.

In a similar kind of way, the burst of the universe is a drop of ink in which all parts are connected. According to Bohm, the elementary particles in the double split experiment are not only connected, but *aware* of each other's existence, bound by the cosmological bond of their primordial origin. The universe, based on the acceptance of the implicate order, is holistic *par excellence*. In this universe all elements are co-dependent on each other, and co-determine each other. In the words of David Bohm:

> It seems necessary to give up the idea that the world can correctly be analysed into distinct parts, and to replace it with the assumption that the entire universe is basically a single indivisible unit. Only in the classical limit can the description in terms of component parts be correctly applied without reservations. Whenever quantum phenomena play a significant role, we shall find that the apparent parts can change in a fundamental way with the passage of time, because of the underlying indivisible connections between them. Thus, we are led to picture the world as an indivisible, but flexible and ever-changing, unit.[11]

The idea of the holistic nature of the universe is not new. What is new is Bohm's justification of it. New ecological consciousness, which has been emerging during the last twenty years, made people aware that nature is one holistic system. Its elements are co-dependent with regard to each other and determine each other. This subtle network of life is of such a nature that if we touch one point of it, the whole network reverberates, as all elements are connected. This we have learnt from the ecology movement and its aftermath.

To what extent David Bohm's cosmological conceptions have been influenced by the thinking of the ecology movement is an open question. It would be fair to guess that there must have been some influence, as the movement had articulated the holistic and co-dependent conception of nature before Bohm's major ideas appeared on the scene. But it is very likely that Bohm may have arrived at his ideas independently. If so, the *Zeitgeist* works through us all. What is important to realize is the fact that the conception of a holistic and co-dependent universe can be rationally upheld both on the scale of nature in ecological habitats, and on the scale of the whole universe. This serves to reinforce many of the conclusions of eco-cosmology we have reached so far.

Our discussion up to this point has centred on the elements of eco-cosmology that can be articulated in cognitive terms, and which draw from the insights and findings of recent science.

The three remaining pillars or principles are related to the ethical order of the human universe – and the interaction of this order with the natural order.

5 The theology of hope

Hope is part of our ontological structure. Hope is a mode of our very being. To be alive is to live in a state of hope. Hope is the scaffolding of our existence. Hope is a reassertion of our belief in the meaning of human life, and in the sense of the universe. Hope is the precondition of all meaning, of all striving, of all action; it is a celebration of awareness. Hope is an essential quality of being human. We are back to our discussion of action. Hope is a precondition of any purposeful action.

At first glance, hope does not seem to be an inherent part of eco-cosmology. But on a deeper analysis it is important for two reasons.

Firstly, hope is indispensable as a force of *continuous transcendence*, thus responsible for the unfolding of evolution on the human level. When our hope crumbles, we crumble. When our hope is asserted, we are alive and participating. Thus, in a subtle way, hope is the will that fuels the participatory mind.

Secondly, and most importantly, hope is a necessity following from the vulnerability of human nature, which needs affirmation, compassion, solidarity, courage and responsibility. The logic of hope *is* the logic of affirmation, compassion, solidarity, courage and responsibility. All these attributes are the very stuff of which human life is made; that is, life that is lived in meaning and harmony.

Can there be a cosmology in which hope does not play any part? Certainly. Such is Sartrean cosmology and such is mechanistic cosmology. However, in cosmologies created by enduring human cultures, hope has always been an essential element of our affirmation of the universe and of ourselves. Eco-cosmology *is* an affirmation of the universe; hope is *part* of this affirmation. Thus hope appears to be an indispensable dimension of those readings of the cosmos in which human life is meaningful in a meaningful universe.

6 Reverence for life

The content of the term 'reverence' possesses an ethical component and a cognitive component; the two cannot be easily separated. Reverence is a consequence of our awareness of the dazzling magic of evolutionary development.[12] When we truly become aware how glorious is the architecture of the universe, how intricate is the tapestry of evolution, how exquisite are the powers of the human mind, and how all these forces of the universe are orchestrated together in one stupendous symphony, our reaction cannot be but of awe and reverence. Thus reverence appears to be an act of an in-depth comprehension – not bits and pieces, here and there, but of the glorious whole, working together.

There is no *logical* necessity to describe the world in reverential terms. Yet for the appreciative and sensitive mind, reverence for life appears as a natural acknowledgement of the miracle and the beauty of life itself. Once we behold the whole universe with reverence, we cannot but embrace the phenomenon of life with reverence. The reverential attitude can then be easily extended to various other species, and to other people; it is, as I shall argue later (see chapters 8 and 9), part and aspect of our overall vision of the cosmos and all its creatures.

Compassion and empathy are the modes of awareness that are known to us all. In some Eastern religions, such as Buddhism and Hinduism, compassion and empathy are not only used as expressions of pity and sympathy, but indeed as modes of knowing. In order to understand another person in depth (and perhaps the same applies to the universe at large) we need to use more than mere abstract, cold intellect. We need to use compassion. *Compassion is reverential understanding.*

It should be immediately clear to us all that what is required for the healing of the planet and for the mending of multiple ecological and social damages is this deeper comprehension, based on the compassion and true recognition of the brotherhood of all beings. Thus reverence, as a form of understanding, becomes a firm pillar of our ecological *Weltanschauung*.

7 Eco-ethics

Eco-ethics is an extension and articulation of the idea of reverence for life. In a subtle way it is also an articulation of the idea of the implicate universe and of the idea of creative evolution.

Let us now look at the whole architecture of eco-cosmology. Although three steps removed, eco-ethics nevertheless partakes in the meaning and the overall purpose of the anthropic principle. In order to proceed with its project of creating life and articulating it, the anthropic principle not only had to create cosmological constants, but had to create evolution as a more specific vehicle for the articulation of life. That much clearly follows. But in a deeper and subtler way it also follows that the participatory mind had to be created if the participatory universe were to make sense of itself through the agency of human intelligence. If the universe did not use this agency, it may have got stuck forever in cognitive darkness and incomprehension. Without the human mind, there would be no light in the universe. Does it sound arrogant? Perhaps it does. But what would the universe be like without the human mind?

Furthermore, in the universe of wholeness and co-dependence, particularly in the universe of human co-dependence and solidarity, specific vehicles had to be created to carry on the heritage of life on the social, ethical and spiritual levels. Reverence for life and ecological ethics are such vehicles.

If life did not create ethical forms of behaviour which safeguard the heritage of life, then the works of the anthropic principle could be in jeopardy on the human level. What I am suggesting is that *eco-ethics is not an invention of weeping ecological softies, but an historical necessity following from the plan of the universe as it unfolds itself from the anthropic principle, via creative evolution, via the participatory mind, to flower in compassion and reverence.* This is a new, intelligent reading of the universe in which reverence and eco-values are envisaged as working partners of the anthropic principle on the level of *Homo sapiens-moralis.* The anthropic principle working intelligently in the age of ecological crisis becomes eco-ethics.

What are the basic values of eco-ethics? Some of them have been already mentioned, if only implicitly. One is reverence for life. Another is responsibility for our own lives, and for the entire cosmos – in so far as we are capable of pursuing it.

The third value is frugality, understood as grace without waste, or as a precondition of inner beauty – frugality conceived not as an abnegation or imposed poverty, but as doing more with less; experiencing life rich in ends, though slender in means. Although it doesn't seem of immediate consequence, frugality, in our individual lifestyles and in our transactions with nature, is terribly important – given the frailty of the earth and the battering it has received. To behave frugally is to show a true solidarity with the planet and its creatures.

Another important value of eco-ethics is the pursuit of wisdom as contrasted with the pursuit of mere information. Yet another is that of self-actualization as contrasted with material consumption. The universe wants us to be wise if only because we can then be intelligent partners and truly appreciate its riches. The universe wants us to pursue the path of self-actualization because only then can we become whole and connected persons and, in consequence, take the responsibility for all there is – becoming guardians, custodians and good shepherds, not plunderers. Eco-ethics clearly follows from a correct understanding of the heritage of life and of evolution.[13]

As we can see, the architecture of eco-cosmology is neither chancy nor whimsical. All its parts are coherently connected together. We are in the midst of a profound knowledge revolution: old patterns of interpretation and perception are bursting like soap bubbles; new pieces of knowledge are crying for integration. This integration is an imperative of our times. The creation of new philosophies that attempt to integrate new explosions of knowledge, philosophies that not only integrate knowledge with the human world, but also attempt to integrate and heal the human being inside; the creation of such philosophies – holistic and assimilative – is a task that is both supremely challenging and supremely important.

My own contribution to our present civilizational debate has been to show: (a) that a new cosmology is indispensable; (b) that it can be rationally justified; (c) that eco-cosmology is coherent and, moreover, it satisfactorily integrates the newest findings of astrophysics with the ethical imperatives of our times.

In brief, eco-cosmology is a small part of a daring effort which aims to provide a new foundation for a civilization that is stuck and that desperately wants to receive a new lease of life. Eco-cosmology provides a backbone, a skeleton of a new world-view. Its justification is a philosophy based on this skeleton. Eco-philosophy, developed in the succeeding chapters of this book, is this kind of justification and a further articulation of eco-cosmology.

A POSTSCRIPT ON ECO-PRAXIS

I have observed that although many people would not recognize themselves as architects of the new cosmology, which I call eco-cosmology, they are in fact actively participating in creating one.

We are all concerned with cleaning the chemical dumps, and many of us are working actively to reduce the hazardous consequences. We are all concerned with reducing air pollution in our cities, and many of us try to do something about it. We were all happy when the ban on the use of DDT was announced – as we didn't relish the idea of being slowly poisoned as the result of it trickling through the food chain. Most of us are quite happy that the ban on cigarette smoking is being gradually enforced in public spaces.

All those acts – of cleaning the air in the cities, of eliminating DDT from the food chain, of eliminating cigarette smoke from our immediate living environments – are acts of eco-praxis. We want to have clean environments and clean air. We want to have non-polluted bodies, free of the toxins that in the long run will produce cancer and other diseases. We want our universe to be clean and unpolluted. Why? What is the deeper reason that makes us desire a clean universe and demand one? What if the universe is a garbage can? What if our bodies are garbage cans? Intuitively, we abhor such notions. Why? Because we somehow *assume* that the universe is not a garbage can. We assume that the universe is a harmonious and coherent place. We further assume that our bodies are wonderful pieces of cosmic machinery:

> What a piece of work is a man! how noble in reason! how infinite in faculty!
>
> (*Hamlet*, William Shakespeare)

All these *assumptions* are related to our larger cosmology. Only if we accept this larger cosmological matrix, which spells for us a positive view of the universe, only then can we justifiably claim that we have the *right* to live in an unpolluted environment. Only then do we have the right to refuse the claim that life is a case of terminal cancer and instead claim that life is a radiant force of enduring beauty.

Through those various acts of eco-praxis (which are practised by many and acknowledged as salutary by all), *we are in fact contributing to the new emergence of eco-cosmology.* The practical implications of eco-cosmology are recognized by all, the cosmological basis from which these practical implications follow is still elusive to many. Yet the two, praxis and cosmology, are connected; they inform each other and subtly, though indirectly, co-define each other.

CONCLUSION

Cosmologies are not arbitrary creations of certain groups of people who fantasize about the cosmos. The creation of a cosmology is a traditional response of a given people to their way of experiencing reality. The experience of reality is never raw, but always mediated by the mind. *The mind of a given people is inherently woven into their cosmology.* As mind allows and guides us, so we build – our lives, our cultures, our cosmologies.

In our day and age, when our consciousness is constantly impinged upon by ecological dilemmas, by the awareness of the frailty of the planet, and of the frailty of our lives in the fragile design, our consciousness is slowly 'greening', and becoming ecological consciousness. This ecological consciousness is a part of our new mind and new sensitivities, which inform us that we must attempt to weave a new pattern of our interactions with nature and with the cosmos. Ecological consciousness represents the interiorization of the principles of eco-cosmology. Cosmological *perestroika* is one part of the design. Ecological consciousness is its symmetrical twin in the sphere of our inner psyche.

Eco-cosmology is a summation of the various attempts that aim at healing the earth by creating a new matrix for our interaction with the cosmos. *Thus the cosmological matrix is never a mere lexicon for reading the cosmos as it is, but is a set of policies for interaction.* In the mechanistic cosmology, this matrix is reduced to control and manipulate, while nature must oblige. In traditional cosmologies, this matrix usually involves a reciprocal give and take, based on participation rather than coercion.

There is a global reconstruction that is going on in various parts of the world aiming at creating a new paradigm. This global reconstruction invariably, though often subconsciously, implies a larger cosmological matrix, which attempts to redefine creatively the multitude of things: the perception of the universe; our reading and description of it; the appropriate modes of acting in it; and last but not least, appropriate ways of treating each other. These four components are reciprocally dependent on each other. If we perceive the cosmos reverentially, this leads us to reverential descriptions, and reverential actions, as well as a reverential treatment of each other. If we perceive the cosmos mechanically, this leads us to mechanistic descriptions, and to a mechanistic treatment of each other. Let me stress that unless we perceive the cosmos reverentially, we cannot hope to act in it with the reverence appropriate for healing rather than harming the earth.

Eco-cosmology, here outlined, is the development of the anthropic principle in the following way: the anthropic principle is the foundation; creative evolution is the temple built on this foundation; we are the choirs singing Gregorian chants in the temple.

The universe gave birth to us – its special form of creation. The universe wants us to succeed. We may consider ourselves special without presuming that we are superior. We are the eyes of the universe. We are the brain cells of the universe. We are the singers singing in the stupendous chapel called the cosmos.

In brief, this chapter has shown that new cosmologies are both possible and inevitable; that new cosmologies can be rational and coherent without necessarily obeying the criteria of scientific rationality. Scientific cosmology is one of many; not the first, and not the best. While creating science, we needed to create scientific (mechanistic) cosmology. In transcending the limits of science, we need to create a post-scientific cosmology. Eco-cosmology is an example of such a cosmology.

Another important conclusion is as follows: from an intelligent reading of the cosmos, an ethics can be derived. Eco-ethics follows from eco-cosmology, and is contained within it. This is how it has been in most known cosmologies: they have engendered their respective ethics – with the exception of the mechanistic cosmolgy, perhaps. But an exception is an exception, not a rule. If some people infer from my argument that the mechanistic cosmology (and empiricist philosophy following it) represents an *un*intelligent reading of the universe, they are absolutely right. This is what I mean to imply.

Eco-cosmology will not be the ultimate design of our philosophical imagination, but in embracing the cosmos in a ferociously participatory and reverential manner, we are testifying to the fact that we are its worthy sons and daughters.

2

Eco-philosophy vis-à-vis *Contemporary Philosophy*

THE DÉBÂCLE OF CONTEMPORARY PHILOSOPHY

Philosophy, like life, is a process of perpetual re-examination, for philosophy is a peculiar distillation of a conscious part of our life. It is an important part of our image of ourselves, which we form in interaction with the external world, with our past history, with our future dreams. Without philosophy, we have no anchor, no direction, no sense of the meaning of life. Each epoch and each society is rooted in some fundamental beliefs and assumptions, which are acted upon as if they were true. They justify all other things that follow from them, while they themselves are accepted on faith. A change in philosophy is a change in the accepted canons of faith, whether that faith is of a religious or a secular character. And conversely, when a given people, society or civilization is shaken or shattered, this calls for fresh thinking; in fact, more often than not, for a new philosophical basis.

It would be commonplace to repeat that our civilization has lost its faith, confidence and direction, and needs a new philosophical basis to get out of its present swamp. It would be commonplace to repeat that past philosophies, including the twentieth-century analytically orientated, Anglo-Saxon philosophy, came into being as the result of a specific distillation of the twentieth-century Western mind, and as such was not only justifiable but perhaps even inevitable. Again, it would be commonplace to observe that when society and civilization take a new turning, philosophy must re-examine its position, shake off the dust of its dogmas and be prepared to be impregnated with new ideas and a new vitality. However, such a process of radical intellectual rethinking cannot be accomplished without some resistance and some pain; for we are all, even philosophers, partial to our dogmas and our mental habits.

Let it be remembered that ours is the age of specialization. And what do we expect of a specialist? That he knows one thing well – even if he is an ignoramus otherwise – that he is thoroughly drilled in this one single thing, and that he is proud of being a narrow technician. In a heaven where the ultimate god is a technician, all the smaller gods are technicians too.

In so far as present-day philosophers stand up to the challenge of the technical age and show their prowess as virtuoso technicians, they are admirable – for technical virtuosity meets with applause in a technical age. In so far as they have had to renounce a part of the great philosophical tradition and drastically narrow the scope and nature of their problems for the sake of the virtuosity, this is deplorable.

But perhaps the philosophers are not to blame; not entirely at any rate. They are simply following the *Zeitgeist*. The whole of civilization has gone topsy-turvy in its zeal for specialization. We are a schizophrenic civilization which deludes itself that it is the greatest that has ever existed, while its people are walking embodiments of misery and anxiety. Our knowledge and philosophy only widen the rift between living and thinking. T. S. Eliot's prophetic cry: 'Where is the Life we have lost in living? Where is the wisdom we have lost in knowledge? Where is the knowledge we have lost in information?' rings today truer than ever.[1]

Philosophers thrive on challenges, for every new philosophy is a challenge *par excellence* thrown to the limits of our comprehension of the world. We are now in yet another period of ferment and turmoil, in which we have to challenge the limits of the analytical and empiricist comprehension of the world just as we must work out a new conceptual and philosophical framework in which a multitude of new social, ethical, ecological, epistemological and ontological problems can be accommodated. The need for a new philosophical framework is felt by nearly everybody. It would be lamentable if professional philosophers were among the last to recognize this. I sense that many of them are diligently groping towards new vistas. Philosophy is a great subject, it has a great past, and a great future. Its present lowly state is an aberration and an insult to its heritage.

Martin Heidegger once remarked that one does not just write books on metaphysics. Metaphysics, and in a sense all philosophy, is a response to the challenge of life, to the challenge of actual problems which are thrust upon us with irresistible force. Genuine metaphysics involves a significant rethinking of the problems of humankind and the world in any given time. In this sense eco-philosophy seeks to provide a new metaphysics for our time. And in this sense the various treatises on metaphysics which analyse only the logical structure of propositions, or attempt to force various levels of being into pre-arranged semantic boxes, are chasing but a shadow of a fading world.

When Wittgenstein proposed his 'logical atomism' it was a genuine metaphysics because it grew out of a real and nagging problem, which was to re-establish solid and coherent foundations for mathematics. It was thought absolutely vital that at least mathematics should be firmly anchored. It was hoped that mathematics, via logic, would provide secure foundations for all other branches of knowledge. Moreover, at the time, the new mathematical logic – which was ingeniously used later as the conceptual backbone of logical atomism – promised to put an end to the chaos of philosophy and also to establish a system of scientific philosophy far superior to anything that had ever existed. Therefore, given the state of knowledge at the time, and given the aspirations of an epoch that still believed in salvation through science, logic, and technological progress, and actually wanted salvation in these terms (one must never discount the aspirations of the epoch as well as the unwritten longings that prompt thinkers and philosophers to move along specific paths), logical atomism was a bold, ingenious, and justifiable venture. Furthermore, given the state of knowledge and the state of minds in the 1920s, Carnap's *Der Logische Aufbau der Welt* (*The Logical Structure of the World*, 1928), was still a legitimate metaphysical proposition, though tending to burst at the seams because it attempted to stitch together too much, too neatly. Perhaps Willard van Orman Quine was the last metaphysician of the epoch – the epoch which sought the resolution of our major problems through logical structures. But today's state of knowledge and today's aspirations do not even remotely resemble those of the 1920s and 1930s. Consequently, anyone still trying to turn philosophy into a neat logical system is chasing a ghost from the past.

Analytical philosophy was born of an aggressive attempt to eliminate all metaphysics and religion. To the degree that the attempt was successful, the consequences were disastrous. The death of metaphysics, so conspicuous among analytical thinkers, proved to be a double-edged sword: for it signified the elimination of many forms of bad philosophy, on the one hand, but it also signified the increasing barrenness of philosophical imagination, on the other. Metaphysics is the fountain-spring of speculative thinking. Without speculative thinking there is no new philosophy.

Indeed, analytical philosophy had so petrified into a set of stale dogmas that it did not allow itself to be renewed by the fountains of inspiration stemming from metaphysical thinking. In killing the weeds in our philosophical garden, we have also killed the bacteria indispensable for growth. The whole field has become sterilized and sterilization has brought about sterility; sterilized fields of thought breed barren minds. The hallmark of all philosophy is *thinking*: opening up new vistas. The capacity for thinking, as distinguished from mere analytical cutting up of language, has been so undermined by the sterile training in present academia that analytical philosophy must face the charge of being guilty of the destruction of thought. Yet so entrenched are analytical philosophers in their role as self-appointed guardians of rationality that in all likelihood they would say that they cannot understand how analytical philosophy, being rational through and through, can be destructive of thought. At this juncture we witness the Freudian shutting-up of gates: analytical philosophers do not want to hear what they do not like to acknowledge. Their rationality closes up their minds to such a degree that they do not listen. At this point there is no dialogue, there is only a dogmatic reiteration of analytical truths and of the sterile dogmas of yester-year. The sterile mind keeps reiterating sterile platitudes. Intolerance towards the views of others can be seen even among the most distinguished of analytical philosophers. Willard van Orman Quine, undoubtedly the most distinguished, still – after so many lessons demonstrating the untenability of logical empiricism – perpetuates the myth of positivism to the point of discouraging students from speculative thought: 'The student who has the propensity to speculative thought is not a good student anyway.'

The final debate is not about empiricism or conventionalism; not even about philosophy; it is about the shape of life. The consequences of Quine's philosophy, and of nearly all analytical philosophy, are belittling the human phenomenon, and degrading and belittling life as it can be lived. For these reasons, analytical philosophy, when it pretends to be the arbiter of our destinies, must be opposed. So often mechanistic philosophies, while heralding the dawn of rationality and the liberation of man from the constraints of dogma, end up as worst oppressors, attempting to bind us with their own dogmas by trying to squeeze our thought into antiseptic, logical boxes. These philosophies also undermine the creative *élan* of the mind and the very rationality they pretend to champion. The paradox is that the more accomplished you become as a master of analytical techniques the less mind you have for the larger philosophical issues. This is not a necessary proposition but a contingent one; its truth is not to be revealed by analytical tools, but only by a deeper reflection into the nature of things. That tools alienate themselves from the original purpose for which they were conceived, is an old story. The tale of the sorcerer's apprentice makes the point very vividly.

Beware of powerful tools. After a while they acquire a life of their own and have a tendency to dominate their inventors. Philosophy was originally conceived as a reflection upon ends; it atrophies when it is reduced to mere means and techniques. As technicians, philosophers do not command any special respect nor do they serve any useful purpose in society. We don't want to be a joke, for philosophy is not a joke but one of the most ambitious and glorious enterprises which humankind has conceived to see the universe more clearly and to live in it more meaningfully.

Let us pay homage where homage is due. Analytical philosophy has done much to liberate us from the spell of language. And Wittgenstein rightly deserves to be hailed as the man who did more than most to liberate us from that spell. But let us recognize that Wittgenstein's philosophy has its own limits, that it is now over forty years since *Philosophical Investigations* was conceived and written, and that since that time we have come to understand that the book constitutes no final tableau of philosophical problems. Indeed, our perspective has changed during the last ten to fifteen years. We have come to recognize that *emergent* philosophical problems are *never* linguistic or analytical in nature. They are part of newly emerging life forms, and as such require adjustments in our ontology and epistemology, in addition to a new conceptual and linguistic apparatus. At present we are once more beginning to be steeped in the problems of the 'real world'. The revision of the entire Wittgensteinian and analytical tradition is in progress.

Now, being conceptually tough and linguistically dextrous (for this is what analytical philosophy best equips us with), we could go on playing linguistic games, and indeed fend off anyone who sought to break our linguistic cocoon; but this would be of no use. For the fact is that by now most philosophers know deep down that a new era of philosophy is coming, that the world expects philosophers to turn their minds to the new philosophical problems of our day, that the linguistic-analytical idiom renders elegant results but is limited in its scope, and that we have been bewitched by Wittgenstein. This is ironic because he warned us not to become bewitched by language itself nor any set of utterances by any particular philosopher.

At the end of his life, the biologist C. H. Waddington, while rethinking the dilemmas inherent in the present state of biological knowledge (which he thought to be mirroring the present state of knowledge at large), blamed philosophy and philosophers for setting us on a wrong trail. He claimed (as have others) that philosophy took a wrong turning at the beginning of the century. Instead of following Whitehead, and his holistic and organistic philosophy, it followed Russell, and his atomist, mathematical philosophy. Waddington's advice was 'Back to Whitehead', which is commendable. However, his diagnosis of the past is fraught with difficulties. For it seems to me that given the entire thrust of the epoch, its belief in progress, exactitude, science, and above all given the conceptual power of the tools of mathematical logic (with its definiteness, crispness, elegance and finality), the search for solutions via logic was too irresistible to allow us to follow Whitehead at an earlier time.

But now it is all different. Logical atomism, logical positivism, the dream of the scientific system of philosophy, and salvation via linguistic hygiene, are all but history. And the present-day world is one in which scientific knowledge is tottering, a world in which the concepts of nature and of ecology are assuming a major philosophical and metaphysical importance. This is a world with unprecedented social and individual afflictions, many of which, paradoxically, have been brought about by a seemingly benign technology which has become our crutch to the extent that we are unable to think and act on our own.

In outlining the scope of what I call eco-philosophy, I shall be pleading for a new system of thought which actually marks a return to the great tradition of philosophy – the tradition that takes upon itself large tasks and attempts to be culturally significant. Since I wish to go beyond the canons and precepts of contemporary philosophy, I cannot be constrained by its criteria of validity. Eco-philosophy, here presented, is offered as a challenge. It possesses enough significant *problems* to make philosophers (and not only philosophers) reflect, ponder, re-examine, propose new insights and truths. Out of creative combats and lively encounters new truths are born; while reciting old truths, whether of Hegelian or analytical philosophy, will only produce a deepening dullness of mind. It is a mark of the enlightened mind to accept the terseness of the challenge with the equanimity of spirit and the generosity of heart characteristic of true searchers for new philosophical horizons. It is my wish that the tenets of eco-philosophy will be received in this spirit.

Let me just add one terminological remark. When I say 'contemporary philosophy' I mean primarily current Western philosophy of the empiricist, analytical, scientific school, since this is the philosophy that not only dominates the Anglo-Saxon universities, but has, indirectly, become the accepted global philosophy. When the Arab countries talk about progress (or an Arab sheikh buys his Mercedes, for that matter), when people talk about development or about bringing education to the illiterate of the Third World countries, it is all done within the implicit context of a Western empiricist, positivist, analytically based philosophy. All major economic transactions in the world are an endorsement of our Western philosophy.

I am aware that there is a difference between logical atomism and Husserlian phenomenology, between the philosophy of the later Wittgenstein and Sartrean existentialism – all Western products. But I am also aware that phenomenology and existentialism have little influence on and do little harm to the world or those private individuals who adhere to their tenets, while empirically orientated positivist philosophy, particularly as developed in the Anglo-Saxon countries, provides the philosophical justification for the ruthless, exploitative, mechanistic paradigm which has wreaked so much havoc on world ecology, on Third World nations and on individuals who have attempted to mould their lives in the image of the machine. And it is this version of contemporary philosophy that eco-philosophy stands against and to which it attempts to provide an alternative.

I am also aware that perhaps none of the present analytical philosophers will recognize himself in my reconstruction. I am not analysing any particular philosopher, nor any set of philosophers, but the essence, and above all the consequences, of the whole philosophical mode of thought of an epoch. Most of all, I am investigating what changes must be made in this mode of thought to make philosophy a truly supportive tool in our quest for meaningful living.

It is precisely this simplistic, linear, atomistic, deterministic – in short, scientific – thinking, that chops everything into small bits and subsequently forces the variety of life into abstract pigeon-holes of factual knowledge which I consider diseased, for in the final reckoning it produces diseased consequences. Therefore, when I say that in devising new tactics for living we shall need to rethink our relationships with the world at large, I mean expressly that we shall need to abandon the mechanistic conception of the world, and replace it with a much broader and richer one. Eco-philosophy attempts to provide the rudiments of this alternative conception.

What is eco-philosophy? How does it differ from what we have agreed to call contemporary philosophy? I distinguish twelve characteristics, which I shall compare with the corresponding characteristics of contemporary philosophy. To heighten the contrast I shall use two diagrams, a mandala of eco-philosophy and a mandala of present philosophy, both of which will be explained in detail as we go along.

THE CHARACTERISTICS OF ECO-PHILOSOPHY

1 Life orientation

Eco-philosophy is life-orientated, as contrasted with contemporary philosophy which is language-orientated. Life is not a 'terminal cancer', as some medical practitioners maintain, but a positive phenomenon with a force and beauty of its own. Those who cannot recognize life's positive vector have already retired from it and allowed themselves to drift into the abyss.

We do not have to justify our partiality to life, for what is more important than taking life seriously? Indeed, the burden of proof is on analytical philosophers. They have to show that their philosophy is of any use to life. We do not intend to be crass about it and ask for some vulgar pragmatic justification of philosophy, or

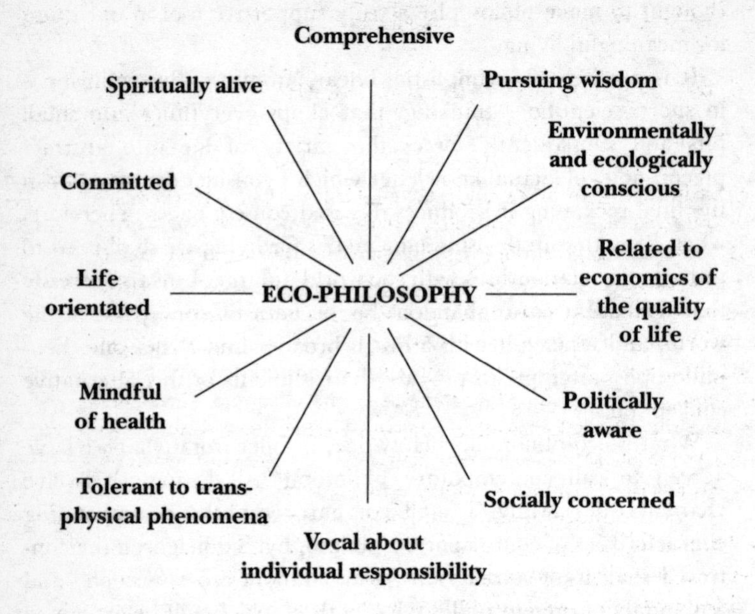

Mandala of eco-philosophy

tell them: 'Show me how your teaching affects my life, or I will fire you.' But some justification must be given in the long run. An obvious one is to suggest that language-rooted philosophy enlarges the scope of our knowledge of both language and the world, and thereby assures enlightenment and provides better tools for living. This 'thereby', however, constitutes a big leap; it is really an article of faith, not a logical conclusion. The whole justification fails if and when we observe that by acting upon this allegedly superior knowledge provided by science and scientifically orientated philosophy, we arrive at major ecological, social and individual pathologies. The point is that in their aloofness, or shall we say in their insularity, academic philosophers so often do not even bother to provide any justification for their philosophy. Philosophy is in the university curriculum. And this is

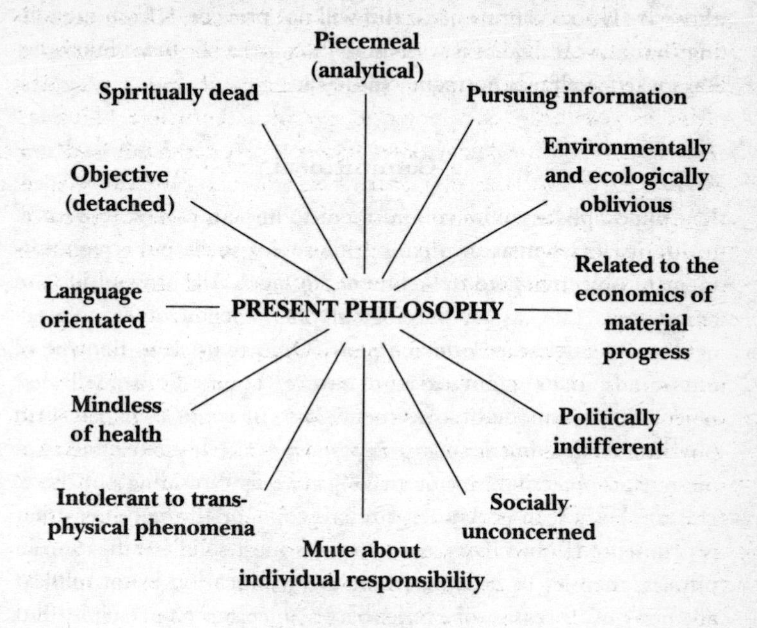

Mandala of present philosophy

good enough for them. However, life has its own ways of avenging.

Philosophy is essentially public and social. Sooner or later, life, through society, or some impertinent individuals, will ask: 'What are you doing, and what is it for?' Sometimes this question is posed to philosophers gently and indirectly, sometimes rather bluntly, as happened at Rockefeller University in 1976, when four distinguished philosophers were fired. So we have no need to be apologetic about maintaining that we want philosophy that is life-enhancing, for all philosophy has only one justification, the enhancement of life. The fact that there is a mountain of self-serving analytical reflection in which so many philosophers are completely buried does not mean anything except that there is this mountain of analytical reflection. We shall not deny that a great deal of brilliance, ingenuity and strenuous effort went into those analytical ventures, but this will not prevent us from suggesting that a great deal of it was energy misspent because philosophy has locked itself in a hermetic cul-de-sac.

2 Commitment

Eco-philosophy signifies commitment to human values, to nature, to life itself, whereas academic philosophy spells out a commitment to objectivity, to detachment, to facts. All forms of life are committed. Life, as an ontological phenomenon, does not recognize objectivity and detachment. Objectivity is a figment of our minds; it does not exist in nature. It can be argued that objectivity is a mode of assessment. If so, it is not rooted in firm physical reality, but is only a *disposition* of the human mind. Let me repeat, objectivity is not an 'objective fact' residing out there. Has anybody seen it? Under a microscope, or through any other instrument? If objectivity is to have its most solid justification in physics, then let us be aware that this justification is not solid at all, not only because of Heisenberg's uncertainty principle, but also because, in the last analysis, we have no way of distinguishing that which is actually *there* from that which our scientific instruments and theories create. In other words, at some point of analysis, when we approach the problem of the existence of ultimate subatomic particles, the (objective) bottom falls out of atomic physics, and we seem to be close to Zen. We co-constitute existence through our perception; the observer is inseparable from the observed.[2]

The concept of objectivity is inseparably linked with the explosion of so-called methodologies, which are, in various disciplines, simply different ways of rendering the same myth of objectivity. The proliferation of methodologies is a menace. Although they were meant to be a tool, a help, in the long run they have become mental crutches, a substitute for thinking.

The fundamental questions 'How to live?', 'What is it all about?' are essentially different from the question 'How to manipulate things?' 'How to live?' belongs to the sphere of eschatology, which is concerned with ultimate goals. Methodologies, on the other hand, belong to the sphere of means which are serving ultimate goals; they are concerned with the ways we handle knowledge.

When eschatology is translated into, and short-changed for, methodology, the question 'How to live?' becomes the question 'How to do things?' This has been one of the tragedies of our times, namely that we have forgotten that there is no methodology that can give an answer to the question 'How to live?' Eco-philosophy maintains that it is a perversion of eschatology to translate it into a methodology.

It is a perversion of the meaning of human life to reduce it to consumption, limiting it to its physical, biological and economic aspects. Ultimate meaning and fulfilment are secured by those singular moments when our being reaches transphysical realms of aesthetic contemplation: of being in love, of deeper illumination when we grasp what it is all about, of religious and semi-religious experiences. All these are transcendental aspects of our being, therefore transphysical and transobjective. We compassionately unite ourselves with the larger flow of life. No philosophy can succeed in the long run unless it attempts to understand nature and life in compassionate terms. Life is a phenomenon of commitment. In avoiding commitment, we are avoiding life. Philosophy that shuns life and a commitment to it is a part of the entropic process leading to death. The death wish of our civilization has pervaded its philosophical edifices. Eco-philosophy attempts to reverse the tide and serve life in a fundamental sense.

3 Spirituality

Eco-philosophy is spiritually alive, whereas most of contemporary philosophy is spiritually dead. I do not have to justify my quest for spirituality in physical terms. If you ask me to do so, it means that you understand nothing of spirituality. It is hard to talk about spirituality to people who pretend that they know nothing about it. Such a posture may be an expression of one's spiritual barrenness. More likely, however, it is a defensive posture: 'I don't like to be manipulated by religious zealots. Therefore, don't even talk to me about spirituality.' Yet who has not been touched by spiritual experience in his or her life?

Spirituality is a subtle subject, difficult to define, and often difficult to defend. Many people reject it because of its traditional associations with institutionalized religion. Let me therefore hasten to reassure you that I am using the term in a newly emergent sense which has little to do with spiritualism, occult practices or established religious connotations. Spirituality, as I see it, is a state of mind – really a state of being. In this state of being we experience the world as if it were endowed with grace, for we ourselves are then endowed with grace. We experience the world as a mysterious, uplifting place. We experience the world in its transphysical or transcendental aspects. The first act of awe when a human was struck with the beauty or wonder of nature was the first spiritual experience. Traditional religions certainly embody this form of experience, but by no means exhaust it. All great art, in its creation and in its reception, is a living expression of human spirituality. Reverence and compassion, love and adoration exemplify different forms of spirituality. The contemplation of great poetry is a spiritual experience *par excellence*.

In order to go beyond our merely biological universe, we as a species have had to refine the structure of our experience, our ability to respond to ever more subtle phenomena, our capacity to experience the world through active intelligence and increasingly versatile sensitivity. Every act of perception and comprehension, when evolution has reached the cultural level, constitutes a subtle transformation of the world. Spirituality is making the physical transphysical. The world experienced spiritually is one in which the process of active transformation through intelligence and through sensitivity is magnified.

Spirituality is, in short, an overall structure for generating our transphysical experience – almost an instrument, enabling us to refine ourselves further and further. Thus, on the one hand, spirituality is a state of being – a peculiar experience of human agents that makes them marvel at the glory of being human or makes them prostrate themselves with compassion or anguish toward other human beings. On the other hand, seen on the evolutionary scale, spirituality is an instrument of the perfectibility of the human. In a sense, spirituality is synonymous with the quality of humanity.

It should be noted that the conception of spirituality I have outlined (although it is independent of traditional religions and treated as a natural phenomenon – an attribute of human existence) does not preclude the recognition of divinity. For in making ourselves into transbiological beings, we needed images and symbols in which our dreams and aspirations could be vested. These images and symbols were deified and institutionalized in various religions, and their presence helped us in our further spiritual journey. Looking at our cultural and spiritual heritage as a whole, we can say without a doubt that the existence of the sacred and the divine has been neither spurious nor incidental but quite essential to the making of the human as a transcendental being.

Eco-philosophy is spiritually alive, for it addresses itself to the ultimate extensions of the phenomenon of man, and these extensions spell out the life of the spirit, without which we are not much more than the chimpanzees jumping from tree to tree. Much of present-day philosophy is spiritually dead, for it addresses itself to problems and realms that systematically exclude the life of the spirit. The language of that philosophy, its concept and its criteria of validity are such that, of necessity, it must dismiss everything to do with spirituality as invalid or incoherent.

To inquire into the condition of the human is to be led inexorably to the conclusion that our essential quest is for meaning. This quest, whether through traditional cultures and religions, or through modern science, is a spiritual quest; it has to do with understanding what it is all about. Thus our essential nature is to try to grasp the stars, even if only to understand where our feet are. To pretend that those ultimate matters are in the private domain of each individual is to give the public domain to the greedy, the rapacious, the exploitative. Great cultures and great civilizations were wiser than that. Some philosophers would maintain that concern with those spiritual matters, important though they are, is not the business of *professional* philosophy. I would maintain that they are mistaken. Philosophy does not have to occupy itself with small and insignificant matters; it has an impressive record in occupying itself with great and important ones. Eco-philosophy has the courage to return to those important matters.

4 Comprehensiveness

Eco-philosophy is comprehensive and global, while contemporary philosophy is piecemeal and analytical. Eco-philosophy is comprehensive not because it is uncritically confident that it can grasp all and explain everything. Far from that. It is comprehensive of necessity, as a result of the realization that we have no choice but to look at the world in a comprehensive, connected and global way. Buckminster Fuller has said that if nature wanted us to specialize, it would equip us with a microscope in one eye and a telescope in the other.

The atomistic and analytical way is one in which, almost of necessity, the trivial, the facile, the obvious and the physical dominate. The ultimate texture of life requires an approach which posits a variety of depths, which assumes that there are things that defy easy analysis (analysis is, in a sense, always easy for it assumes that things must fit the tools with which we approach them) and which also acknowledges that these are the things that ultimately matter. All eschatology is non-analytical.

Eco-philosophy, perceived as global and comprehensive, is a process philosophy which is integrative, hierarchical and normative – self-actualizing with regard to the individual, and symbiotic with regard to the cosmos.

One of the most delicate matters of knowledge and philosophy is that of truth. Eco-philosophy believes that truth is a far more intricate affair than simply finding an adequate description for our facts within the frame of reference of physical science. We acknowledge that truth consists of a correspondence between reality and its description. The notion of reality cannot be simply exhausted by scientific frames of reference however.

As we all know, in ecology we assume a much broader frame of reference than in physics or chemistry. Consequently, a mere physical or chemical description of phenomena, when we are in the ecological frame, will not do. But ecology is not the ultimate frame of reference. Evolution provides a much broader frame, particularly when it includes the cultural evolution of the human. So our concept of truth must be related to the frame of evolution at large – not to a static description of things within 'evolutionary theory', as provided by molecular biology, but within evolution unfolding, evolving, producing newly emergent forms. Ultimately it makes sense to relate the concept of truth to the *cosmic scale*, within which evolution occurs. However, there is a problem here: one has to be omniscient to grasp the place of particular phenomena in evolution occurring within the cosmic scale. Therefore, we must be exceedingly cautious when we handle truth, for so much depends on our adequately describing our concept of reality. Perhaps it would be closest to truth to say that all claims to it are approximations because there is only one truth about everything. Such a conclusion is not going to be comfortable to the mind which is used to rigid categorizations and to the ascription of truth to single statements. We know how constrained the physical frame of reality is, and how constrained are its 'truths'. The cosmic scale is harder to grasp and harder to live with. However, we should not be concerned with making things easy but with understanding their ultimate reality.

5 The pursuit of wisdom

Eco-philosophy is concerned with wisdom whereas most existing philosophies are directed towards the acquisition of information. It is not easy to talk about wisdom without sounding pretentious. What is wisdom? Even the wise are at a loss to answer this question. Wisdom consists in the exercise of judgement, based on qualitative criteria, usually in conflicting situations. Judgement cannot be quantified, neither can compassion, which is often a part of judgement. Thus wisdom, too, is essentially unquantifiable; it is an embarrassment to the quantitative society as it defies its very ethos; but at the same time, paradoxically, it is a quality highly sought after as it is recognized that fact and measurement can carry us only so far.

The influence of our present quantity-ridden society and our present quantity-ridden education – one is the mirror-image of the other – is so pervasive that we are positively discouraged from exercising judgement and are prompted to make decisions 'on the basis of facts'. 'Facts are not judgemental; facts do not judge', we are told. But there is a huge fallacy in this proposition, for in a subtle way facts do judge; facts are judgemental. *To obey facts is to obey the theory and the world-view which those facts serve and which they exemplify and articulate.* Facts are thus imperious judgements on behalf of the emperor called the Physical Paradigm of Reality. There is no escape from judgements – even when we accept the judgement of facts.

Wisdom is the possession of *right* knowledge. Right knowledge must be based on a proper understanding of the structural hierarchies within which life cycles and human cycles are nested and nurtured. E. F. Schumacher writes: 'Wisdom demands a new orientation of science and technology towards the organic, the gentle, the non-violent, the elegant and beautiful.'[3] Ultimately, wisdom must be related to our understanding of the awesome and fragile fabric of life. For this reason alone it must entail compassion, for compassion, properly understood, is one of the attributes of our knowledge of the world. It is a crippled school in which compassion and judgement are not developed. It is a crippled society in which judgement and compassion are neglected, for they are essential to acquiring some rudiments of wisdom – without which life is like a vessel without a keel. (More about wisdom in chapter 5.)

6 Environmental and ecological consciousness

Eco-philosophy is environmentally and ecologically conscious, whereas contemporary academic philosophy is very largely oblivious to environmental and ecological concerns. It is of course so by definition, though there is a great deal more to eco-philosophy than simply caring about our natural resources. Being ecologically conscious not only means taking judicious stock of existing resources and advocating stringent measures to make them last longer; it also entails reverence for nature and a realization that we are an extension of nature and nature an extension of us. Human values must be seen as part of a larger spectrum in which nature participates and which nature co-defines.

It may be argued that it is unfair, and indeed far-fetched, to accuse contemporary philosophy of lack of concern with ecology when it is simply silent on the subject. This is precisely the point: by its silence it participates in the conspiracy of indifference. Crimes of silence are particularly reprehensible in those who ought to be aware. Besides, contemporary philosophy indirectly endorses the view that everything is a matter for specialists and that, therefore, questions concerning the environment and ecology are to be left to the specialists, to economists, politicians, engineers, managers. Any philosophy worthy of the name must perceive that our views on ecology and the environment are always pregnant with eschatological, philosophical and ethical consequences.

7 Economics of the quality of life

Eco-philosophy is aligned with the economics of the quality of life. Academic philosophies in the West seem to be unrelated to *any* economics but are in fact aligned with the economics of material growth, for they function within a framework which not only tacitly supports but in fact engenders the ideal of material growth.

Western academic philosophers are empiricists or at least deeply affected by empiricism. They adhere, by and large, to the secular world-view, recognize in material progress a valid measure of progress (and perhaps the sole definition of progress), and therefore clearly, though indirectly, support the modus of the economics of growth. The simple fact is that empiricism provides a philosophical justification for the economics of material progress.

Empiricism, material progress and the economics of growth are all intrinsic parts of secularism conceived as a world-view. Empiricism explains the world as being made of empirical stuff, material progress postulates that human fulfilment has to do with material gratification, while the economics of growth is the vehicle which secures the goods desired by material progress. There is no justification for the economics of growth in itself; its *raison d'être* is that it fulfils the requirements of material progress. Thus empiricism is the root, material progress the trunk and branches, and the economics of growth the fruit of the tree of secularism.

Eco-philosophy believes that an economics which undermines the quality of life is in conflict with life itself. Hazel Henderson, E. J. Mishan, E. F. Schumacher and others have shown the fatuity and meaninglessness of an economics geared to material growth alone.

The forces that determine the future of society, and of the individuals living in it, cannot be a matter of indifference to the philosopher. For this reason an understanding of economics, in terms of its relationships with nature and in terms of its influence on present society, is certainly a philosophical undertaking.[4]

8 Political awareness

Eco-philosophy is politically aware; it is also politically committed but not in a superficial way, however. Eco-philosophy is political in the Aristotelian sense: we are political animals not because we crave power, but because our actions are pregnant with political consequences. In short, *we make political statements not so much by the way we vote as by the way we live.*

Take one specific and rather drastic example. The population of the United States produces over 360 million tons of garbage per year, which means 1.8 tons per year per person, or 10 pounds a day. No other country can even begin to approximate this feat. To shovel away this pile of garbage (which, according to one estimate, is enough to 'fill 5 million trailer trucks, which if placed end to end would stretch around the world twice'), American taxpayers contribute $3.7 billion a year towards garbage disposal. Compare this with some other annual spending figures: urban transport, $130 million; urban renewal, $1.5 billion a year.[5]

There is a clear political statement involved in this production of garbage. In participating in it one is participating in an orgy of waste, with all the consequences. One of those consequences is a peculiar kind of mental pollution: getting used to waste as a way of life. Now, in order that America can waste, other nations must contribute. And do they contribute on their terms? No, they contribute on America's terms. Why? Because in this technological world the suppliers provide their goods on the terms fixed by the consumers. And the result of it? Quite often gross injustices and inequities. The plight of the Bolivian peasant or the Brazilian plantation worker, indeed the plight of most Third World manual workers, is directly linked to the way the industrial countries (the consumers) choose to conduct their affairs. The production of garbage is ultimately a political act through which we (indirectly) affect the lives of others. The equation, alas, is simple: the more garbage we produce (and, in general, the more we consume), the more adversely we affect other people who are at the supplying end. Looking around, one can clearly see that political structures and alliances are forged and maintained (sometimes with far-reaching and not always pleasant consequences for local populations) in such a way that oil and other natural resources can flow to the industrialized countries.[6]

9 Social responsibility

Eco-philosophy is vitally concerned with the well-being of society. It regards society as an entity *sui generis* which possesses a life of its own. Consequently, society can neither be reduced to individuals (or considered as a mere sum total of particular individuals), nor can it be understood through its 'outward behaviour'. Society is the nexus and cradle of aspirations and visions which are certainly transindividual. Society is ultimately one of the modes of our spiritual being, and is certainly many other things too: an instrument for transacting business, an insensitive bureaucratic beast that frustrates our quest for meaning. But ultimately it must be viewed as an instrument of our perfectibility; thus, in the metaphysical sense, a mode of our spiritual being.

The social contract by which we are bound is cooperative by its very nature; it is only an acknowledgement of our belonging to the larger scheme of things called the cosmos. It is quite clear that a compassionate, symbiotic and cooperative conception of the cosmos, of necessity, implies a cooperative conception of society, for society is one of the cells of the cosmos in its evolution.

Academic philosophy includes as one of its components social philosophy. But within its scope society is treated as if it were an insect under the microscope: it is all analytical scrutiny, with little concern for the well-being of society. It is no accident that many contemporary philosophers regard society as a mechanistic aggregate to be handled in terms of observable behaviour and by means of statistical laws. Such a treatment does gross injustice to us as social and human beings.

10 Individual responsibility

Eco-philosophy is vocal about individual responsibility. It insists that in addition to the rights we crave, we are also bound by duties and obligations. The point has been made by Solzhenitsyn: 'The defence of individual rights has reached such extremes as to make society as a whole defenceless against certain individuals. It is time, in the West, to defend not so much human rights as human obligations.'[7] But eco-philosophy also observes that the sovereignty and autonomy of the individual must be restored so that he can exercise his rights and responsibilities meaningfully.

The world of the specialist is a world in which all sorts of crutches progressively supplant our limbs and other organs, including the mind; it is a world in which our will and imagination are slowly replaced by mechanical devices; our initiative by the central computer. There is no doubt that part of our crisis is a crisis of confidence, which is in direct proportion to our delegation of powers to the expert, the specialist, the machine. And a great deal of the violence in our society is the result of our frustrated quest for responsibility and initiative. Unable to do significant things on our own, we find an outlet for this frustrated quest in pathological forms of behaviour: violence, destruction, rape. (Rape, on one level of analysis, is a vengeance against society.)

Eco-philosophy suggests and insists that *we* are responsible for everything, including the possibly fantastic transformation of the world to a degree approaching omega point, the end of time, at which man fulfils his destiny by becoming God.[8] Eco-philosophy believes in the human will as a manifestation of the divine. We are the new Prometheans who have the courage to light the fire of our imagination *de novo*; but we are also aware of hubris, and of the enormous responsibility that the carrying of the flaming torch entails.

11 Tolerance of the transphysical

Eco-philosophy is tolerant of transphysical phenomena. The desire to understand the cosmos is as deeply rooted in our nature as our impulse to survive in physical terms. Knowledge is therefore not only an instrument of survival, but above all the ladder that we climb to reach the heavens. We live all the time in a multitude of webs signifying different orders of being and spelling out the complexity of our relationships with the world. In this multitude, the physical web is just one. However, it is this particular web which has become the focus of our attention and the object of intense investigation. We have become so obsessed with it that we have nearly lost sight of all the other webs, although those other webs are always present. We know it. We are using a different sense of 'know', however, than the one that is officially acknowledged. We have great difficulty in expressing, in current language, this different sense of 'know', for current language has become monopolized, and in a sense perverted, by the physical web. Language is always important. Be aware of that!

Eco-philosophy terminates this monopoly, as it calls for a pluralistic epistemology designed to investigate orders of being and orders of knowledge which are both physical and transphysical. To transcend physics and go beyond its universe is the kernel of all philosophy, for the term metaphysics stems precisely from the desire to go beyond physics. One of the basic preoccupations of philosophy through the millennia has been an attempt to penetrate orders of being beyond the physical.

Although our enterprise is ontological and cosmological, as we try to determine and map the heterogeneity of the universe and our relationships with it, our *problem*, at present, is *epistemological*. For there is a peculiar monopoly in epistemology which we have to break in order to be able to talk about other orders of being. If we do not do that, we shall be rendered speechless by the proponents of the present epistemology, be they philosophers or scientists, who will invariably ask: how can you *justify* your claim, what is your evidence for it? By 'justify' they mean physical justification, in 'accepted terms', within the framework of accepted empiricist epistemology and its various tributaries called methodologies. Thus 'justified claims' bring us back to the one-dimensional empiricist universe. So, if we are to arrive at a pluralistic epistemology, we must go beyond this constraint.

Can you justify acupuncture? You simply cannot; that is, if by justification you mean a satisfactory explanation of the phenomenon in the currently accepted empiricist frame of reference. Also, how can you justify the reservoir of biological knowledge that we all possess, on which we vitally depend, and which we indirectly acknowledge when we refer to our instinct, cunning, prescience, premonition, insight, wisdom, compassion? Can you justify telepathy, clairvoyance and other paranormal phenomena? You cannot. But you cannot dismiss these phenomena any longer with the exclamation ''tis all a quackery'. Philip Toynbee writes:

> One of the most depressing aspects of the whole affair [the investigation of paranormal phenomena] is that – at least during the last seventy years of serious investigation – the scientific establishment has wallowed in rancorous and punitive obscurantism which is truly reminiscent of the Inquisition.[9]

Eco-philosophy signals the beginning of a new epistemology: pluralistic, life-rooted, cosmos-orientated in contradistinction to the present one which is matter-rooted and mechanism-orientated. One point must be firmly borne in mind: a great deal of present philosophy, particularly the analytically orientated, consists of mere footnotes to the empiricist epistemology. That epistemology, remember, indirectly represents a constraining universe conceived in the image of a deterministic machine. So let us not get caught in the toils of present epistemology and its various methodologies with their criteria of *justification*, *evidence* and *validity*, for they are all a part of the Cognitive Mafia, guarding the monopoly of the one-dimensional objectivist physical universe. These methodologies are but ornaments engraved on a tomb; they have nothing to do with life and with the epistemology of life. Eco-philosophy insists that in the long run we must create the epistemology of life. The task now is to clear the rubble from the ground and expose the limitations of contemporary philosophy in so far as it has become a deferential tool perpetuating a crippled and crippling conception of the universe.

12 Health consciousness

Eco-philosophy is health-conscious, whereas most schools of contemporary philosophy ignore this question. We are physical aggregates in motion, but we are also luminous chandeliers emanating thoughts, emotions, compassion. Eco-philosophy abolishes the Cartesian dualism of mind and matter and regards the various states (or orders) of being as parts of the same physico-mental-spiritual spectrum. The whole story of the universe is of matter acquiring sensitivity – to the point of awareness, to the point of consciousness, to the point of self-consciousness, to the point of spirituality. Reason itself is a form of sensitivity of matter. This whole physico-mental-spiritual spectrum is our responsibility, and maintaining our health is our responsibility. We are not machines to be mended when one part is broken or worn out; we are exquisitely complex fields of forces. Only when we assume that humans and the environment are made of fields of interacting forces do we begin to understand what a fascinating story the maintenance of human health is, and how miraculous it is when things are in order, and we are in a state of positive health. To keep this field of forces in constant equilibrium means being in touch with the variety of transphysical forces which contribute to that equilibrium. *To be in a state of positive health is to be on good terms with the cosmos.*

New thinking about health is slipping through even to the heart of the establishment. Thus, John Knowles, President of the Rockefeller Foundation, writes in a 1978 issue of *Science*:

> Prevention of disease means forsaking the bad habits which many people enjoy . . . or, put another way, it means doing things which require special effort – exercising regularly, improving nutrition, etc. . . . The idea of individual responsibility flies in the face of American history, which has seen a people steadfastly sanctifying individual freedom while progressively narrowing it through the development of the beneficent state . . . the idea of individual responsibility has given way to that of individual rights – or demands, to be guaranteed by government and delivered by public and private institutions. The cost of private excess is now a national, not an individual responsibility. This is justified as individual freedom – but one man's freedom in health is another man's shackle in taxes and insurance premiums. *I believe the idea of a 'right' to health*

> *should be replaced by that of a moral obligation to preserve one's own health* [my italics, H.S.].

Why should this concern with one's health be elevated to a philosophical proposition, when every boy and girl in elementary school is told: 'Take care of your health'? Within eco-philosophy, taking care of one's health means taking responsibility for the fragment of the universe which is closest to one, expressing a reverence towards life through oneself; it is part of the new tactics for living.

An aspect of our responsibility for our total health, or perhaps even its precondition, is recognition of the sanctity of life. The sanctity of life is not something you can prove with the aid of science. The sanctity of life is an assumption about the nature of life, particularly as perceived, comprehended and experienced by human beings. Recognition of its sanctity appears to be a necessary prerequisite for the preservation of a life worth living. Now, if I experience life as being endowed with spirituality and sanctity, who are you to dismiss my experience with a few titbits of empirical data? It is no good arguing that 'science does not lend any support to the supposition of the sanctity of life', for, in a sense, science does nothing. It is people, enlightened or unenlightened, rapacious or compassionate, who use science to support their views, opinions and visions. But there is an issue here. Our perception and comprehension are carried out, made valid and meaningful, within a conceptual framework. The one based on science seems to preclude recognition of the sanctity of life. But *this conceptual framework itself is a form of mythology.* In insisting on the sanctity of life we are clearly operating in another conceptual framework.

All world-views, like all civilizations, are ultimately rooted in mythologies. I am using the term 'mythology' not to signify a fable or a fiction but rather a set of assumptions and beliefs which form the basis of our comprehension of the world. The ancient Greeks had their colourful mythology. Medieval Europe had its religious mythology. All so-called primitive societies had their respective mythologies. For all its claims to the contrary, science is a form of mythology. It has its unwritten and unproven dogmas which are otherwise called the presuppositions on which science is based. It accepts uncritically and unapologetically a form of voodoo, otherwise known as scientific method. It worships certain deities, otherwise known as objective facts. It deifies certain modes of behaviour, otherwise known as the pursuit of objectivity. It gives sanction to a certain moral order, otherwise known as neutrality.

As in classical mythologies, all these characteristics are interconnected and interdependent. Neutrality is a necessary moral ingredient to make the pursuit of objectivity a privileged, preferred, superior mode of behaviour. Objectivity is, in turn, necessary for making 'objective facts' our deities. Making objective facts our deities in turn justifies scientific method, which is so conceived as to enable us to explore and enshrine precisely those kinds of facts. Objective facts and scientific method are, in turn, necessary to 'justify' the presuppositions of science, for those presuppositions are so conceived as to reveal to us only what scientific method allows for, in other words, what is contained in the notion of physical facts. The structure of the scientific mythology is no less complex than that of traditional mythologies, and no less self-defining.

I am neither deriding nor trying to diminish the importance of science. Mythologies are of major importance in the life of societies and civilizations. We cannot readily perceive that science is a form of mythology because science is the filter or telescope through which we interpret the world. When we use it, we perceive what it reveals; but very rarely what it is itself.

Besides, tampering with science and its mythology means tampering with the whole reality science has constructed for us. We are reluctant to tamper with our basic view of 'reality' for this would create too great a challenge to our identity – which is partly formed by the scientific view of the world. We cling tenaciously to the mythology of science for so much of it was poured into us at school when we were of a tender, uncritical age. We cannot successfully challenge it, or liberate ourselves from it, unless we develop an alternative mythology. The creation of an alternative world-view or an alternative mythology is the imperative of our times. Eco-philosophy offers itself as a possible candidate.

To sum up, let me emphasize that the first diagram (p. 40) is not a catalogue of the virtues of eco-philosophy here proposed, but a graphic representation of the overall belief that until and unless we acquire a conceptual scheme (call it philosophy, if you will) that is comprehensive and encompassing enough, we shall not be able to accommodate and articulate the variety of new relationships which are necessary for an ecologically healthy and humanly harmonious world view.

Let us also notice the essential interconnectedness of the two diagrams. Each signifies a totally different paradigm. When we move around the individual components of each diagram, we notice that each component, in a subtle way, determines the next, and is itself subtly determined by the previous one. Contemporary philosophy cannot help being spiritually dead, for its universe is dead: inanimate matter, physical facts, objective logical relationships. For this reason, having at its disposal the concepts that are specific to this dead universe, it cannot help being socially unconcerned, for social concern is not an objective category. It cannot help being politically indifferent, for politics is too large for its scope. It cannot help being mute about individual responsibility, for the idea of responsibility is beyond its scope and jurisdiction. It cannot help pursuing information, for information consists in those bits that perfectly fit its requirements, while wisdom does not. It cannot help being environmentally and ecologically oblivious, for its hidden premise is that the environment is there to be mastered by man and exploited to his advantage. It cannot help supporting, if only indirectly, the pursuit of material progress. It cannot help being oblivious to health for, according to it, health is the province of the medical specialists. It cannot help being intolerant of, if not hostile to, transphysical phenomena, for they violate the universe of its discourse which it takes to be valid and immutable. Behind the crippling narrowness of academic philosophy looms the shadow of logical empiricism (with its conception of pseudo-problems) which was used as a hatchet to eliminate from the domain of philosophy the most significant and vital problems.

Now, if we start from a different cardinal premise, for example that philosophy is life-orientated and that its mission is the enhancement of life, then all the other characteristics of the first diagram, that is of eco-philosophy, follow. The new philosophy must be spiritually alive in order to understand the human being, a spiritual agent. It must concern itself with wisdom, for we do not live by physical facts alone. It must be ecologically concerned and support the economics of the quality of life. Let me underscore some of the main conclusions of eco-philosophy rather than reiterate its characteristics: objectivity does not exist in nature; wisdom is essentially unquantifiable; life not based on qualitative criteria is meaningless; we make political statements not so much by the way we vote but by the way we live; society is one of the modes of our spiritual being. Pluralistic epistemology is tolerant of transphysical phenomena and embraces a variety of modes of being.

In his book, *A Guide for the Perplexed*, E. F. Schumacher maintained that one of the most urgent tasks of our times is a metaphysical reconstruction. Once we know what we are doing and *why*, other forms of reconstruction, including the economic one, will follow more swiftly. For it is undeniably the case that if our foundations are cracking, no partial reconstruction or repair at the top of our edifice will be of any avail. A number of writers have explicitly endorsed Schumacher's programme and attempted to provide some parts of this reconstruction. But while their works have a more practical and economic orientation, I address myself to the very foundations, to the philosophical and value problems which lie at the core of our intended metaphysical reconstruction.

CONCLUSION

This, then, is the essential message of eco-philosophy: we can affect every element of our social, individual, spiritual, ecological and political life, not separately, but by affecting them all at once. Moreover, unless we affect them all, none will be affected. This is at least a partial explanation of why so many excellent alternative schemes (like the ecology movement) seem to me to have failed. Their vision was too limited. They addressed themselves only to a part of our mandala and regarded that part as the whole.

Eco-philosophy is another chapter in our continuous dialogue with the ever-changing universe. In changing ourselves and our relationships with it, we are changing and co-creating the universe. Out of the lethargic trance of technological inertia, we are emerging with a heightened awareness of our destiny, which is to build a responsible world by assuming our own responsibility, to infuse the world with meaning and compassion, and to carry on the unfinished Promethean story: the story of man unfolding – of which great systems of past philosophy are such a luminous and inspiring example.

I am aware that at times my analyses may seem a bit belaboured. The point is that we must be able to anticipate and meet the arguments of the proponents of the status quo, who have monopolized the language of our discourse and, deeper down, our epistemologies. A dismantling of the entire epistemological strait-jacket (which is linked with a particular world-view) is our road to liberation. We start our liberation with liberating our minds.

In the discussion of eco-cosmology (chapter 1) I showed that there is a connection:

$$\text{Cosmology} \rightleftarrows \text{Philosophy} \rightleftarrows \text{Values} \rightleftarrows \text{Action}$$

In the present chapter I have articulated the second link of this connection – philosophy. In the next chapter I shall articulate the third link – values.

3

Knowledge and Values

Let us begin with certain distinctions that are fundamental to the scientific world-view and are at the same time responsible for many of our present problems, conceptual and otherwise. One is the distinction between knowledge and values. The separation of these two was a momentous event in the intellectual history of the West, leading as it did to the emancipation of specialized scientific disciplines from the body of natural philosophy. But it was a perilous event, too, in that it led in the long run to a conception of the universe as a clock-like mechanism and to the gradual elimination of these elements of our knowledge which disagreed with that mechanistic view – including intrinsic values, which were replaced by instrumental values.

Logically there would seem to have been two different processes involved: intense exploration of the physical world on the one hand and the slow disappearance of intrinsic human values on the other. This logical separation is misleading, however, for what we witness here are two aspects of the same process. The quest for scientific explanations and the growth in importance of the physical sciences coincided with, indeed took place in the context of a decline in the importance of intrinsic values. Our vast store of knowledge of the physical world can thus be said to have been accumulated at the expense of human values. This is a large claim and the present chapter will attempt to justify it. Also in this chapter I shall argue that there appears to be a see-saw relationship between factual knowledge and intrinsic human values: as one goes up, the other is pushed down. If this perception is correct, then it would follow that the resurrection of intrinsic values and their reinstatement at the centre of our lives may indeed come about but that it will be at the expense of our adulation of science and of the physical fact, which we have exaggeratedly elevated to the status of deities.

BASIC HISTORICAL POSITIONS

Historically we can distinguish at least four basic positions regarding the relation of values to knowledge.

The first is the position of classical antiquity as exemplified by Plato: values and knowledge are fused together; one does not become dominant or subservient to the other. As we know, Plato believed in the unity of truth, goodness and beauty. Within his universe values and knowledge are two aspects of the same thing; no knowledge is value-free, and no values can be regarded as void of knowledge. According to Plato, to possess superior knowledge is to lead a superior life. Knowledge is a vital part of the network of life. Most sins are the fruit of ignorance.

In the Middle Ages we can distinguish a second position: knowledge is in the service of values, but at the same time it is subordinated to values which are determined by the Church. Knowledge is fused with values and must agree with values that are accepted a priori as supreme. To grasp God's design, God's order, and the values that follow from that order sometimes required faculties stronger than the mere human intellect, which at times saw discrepancies between natural reason and God's order. Hence revelation was accepted as a mode of cognition, for it allowed one to transcend reason and to find a justification for the fusion of knowledge and values under the supremacy of values.

The remaining two positions can be clearly discerned *in the post-Renaissance period.* The third position separates knowledge from values, without, however, giving supremacy to either. This position is perhaps best represented by Immanuel Kant (1720–1804), who clearly saw in Newtonian physics indubitable knowledge governing the behaviour of the physical universe – a separate realm unto itself; but who, at the same time, would not subject the autonomy and sovereignty of man to any deterministic set of physical laws. Hence he summarized the autonomy of both realms by declaring: 'The starry heavens above you and the moral law within you.'

The fourth position is, of course, *the one held by classical empiricism* and its more recent extensions: nineteenth-century positivism and twentieth-century logical empiricism. This position separates values from knowledge and, by attaching supreme importance to knowledge of things physical and by ruling that values are not proper knowledge, it *ipso facto* establishes the primacy of knowledge over values. This tradition is so near to us and envelops us so constantly and consistently that we are often unable to see through it in order to assess its impact on us.

Thus the four basic positions are:

- Plato – the fusion of knowledge with values without asserting the primacy of one over the other.
- Christianity – fusion of the two but asserting the primacy of values.
- Kant – separation of the two without censure of either.
- Empiricism – the separation of values from knowledge while asserting the primacy of (factual) knowledge over values.

It is, of course, the empiricist position, or the empiricist tradition, that we want to examine in some detail, for this is the tradition that looms largest on our intellectual horizon; this is the tradition that has become our intellectual orthodoxy, the tradition that has been programmed into our ways of thinking and judging, the tradition that has brought the value-vacuum to our society, to our universities, to our individual lives.

The life of cultures and societies is an exceedingly complex affair. What we must do is to unravel the multitude of causes and effects and then see how the original visions and insights of Bacon, Galileo, Descartes, etc. have given rise to larger doctrines, been channelled into various tributaries of learning and life, and been reinforced in the process; and how the process continues to feed upon itself by outlining the boundaries of its territory and maintaining rigid control over what is legitimate within the territory and what is illegitimate. To give two specific examples: research into chemical warfare is 'legitimate', for it is an extension of 'objective knowledge' into the sphere of 'certain chemicals'; research into acupuncture is 'illegitimate', because the phenomenon itself seems to undermine certain fundamental tenets of the empiricist world-view. The connection between a particular phenomenon, or a particular strategy, and the basic tenets of the world-view is indirect and is usually several steps removed, but it is there, if we have the patience and perseverance to look for it.

Strange as it may seem, this connection is often more readily grasped by intellectually 'unsophisticated', rebellious youth than by the 'sophisticated' minds that govern present-day academia. It is remarkable that, on the basis of some inner moral feedback, young people can sometimes react with strong moral revulsion and complete moral conviction to abuses of knowledge in academia and elsewhere, while academia itself often seems oblivious to those abuses.

The intellectual tradition that has directly and indirectly caused the value-vacuum has its roots in the seventeenth century, during which time the doctrines of Bacon, Descartes, Galileo, Newton, Hobbes, Locke, Hume and others were remoulding the world, or rather our picture of it, to make it independent of religion. In the eighteenth century the centre of gravity moved to France, where d'Alembert, Condillac, Condorcet, Diderot, Voltaire, Laplace, La Mettrie and others furthered the cause of secularism and of the scientific world-view. Then in the nineteenth century the tradition was continued by Auguste Comte in France, Jeremy Bentham and John Stuart Mill in Britain, and by the leading materialists: Feuerbach, Marx, Engels and Lenin. In the twentieth century the tradition was further articulated, refined, and couched in more sophisticated language by Bertrand Russell in Britain and by the logical empiricists of the Vienna Circle.

More recently this tradition has found its extension in analytical philosophy, in behaviourist psychology, in operationalized social science, in quantity-ridden and computer-obsessed political science, and in quite a variety of other disciplines, which are full of facts and figures even if those facts and figures explain precious little.

I have sketched the line from Francis Bacon to B. F. Skinner as if it were one uninterrupted, homogenous development; as if the present predicament were the result of some inexorable logical process. The process was far from homogenous. What is really startling is the fact that, in spite of a great variety of opposing intellectual forces, the scientific-empiricist world-view has prevailed so remarkably.

Parallel to the prevalent empiricist tradition there ran, and still does run, the other tradition, which for lack of a better term we shall call anti-empiricist. This tradition was represented by minds at least as powerful and superlative as those on the empiricist side: Pascal, Leibniz and Spinoza in the seventeenth century, Rousseau and Kant in the eighteenth century, Hegel and Nietzsche in the nineteenth century were all seeking a world liberated from the constraints of scholastic theology, but which was not reducible to quantity and measurement.

Pascal's case is particularly illuminating, for he, more clearly than perhaps anyone else in the seventeenth century, saw the great value and the great attraction of science and, at the same time, the great danger in unconditional submission to science. He wrote: 'Knowledge of physical science will not console me for ignorance of morality in time of affliction, but knowledge of morality will always console me for ignorance of physical science.'[1]

Equally illuminating is Spinoza's case. His *Ethics – Demonstrated in the Geometric Order* is the work in which he argues that the good is everything which furthers knowledge, and vice versa. Happiness consists solely in knowledge. Virtue itself is knowledge. 'Happiness is not a reward for virtue, but virtue itself.' He further argues that love can be conceived as the perfectibility of the human through knowledge, for knowledge induces love – a position not far removed from Plato's. What is most curious about Spinoza's *Ethics* is that it attempts to prove its propositions as if it were a textbook of geometry. Though departing radically from the scientific tradition which was later to prevail, Spinoza paid lip service to it (and more than that) in attempting to provide geometrical (scientific?) demonstrations of his ethical convictions.

In the eighteenth century Rousseau and Kant defended, in their respective ways, the autonomy of the human world against the encroachment of the mechanistic world-view and the spreading wave of empiricism. Of the two, Rousseau was the flamboyant one, while Kant was the incisive one. Rousseau eloquently, and sometimes dramatically, protested against 'civilization', which he thought estranged us from our essence and from our fellow humans. The 'artificial' ways that civilization imposes on us are at the source of individual and social alienation. This was a prelude to twentieth-century outcries against science and technology for imposing on us their artificial ways.

Kant, on the other hand, held that if empiricism is correct, we possess no certain knowledge of the physical world; if we do possess such knowledge, in the laws of physics, then empiricism (which insists that the sources of this indubitable knowledge are the senses) collapses. Kant felt compelled to conclude that knowledge of physics provides only a knowledge of the appearances of things, not of 'things-in-themselves'. He held, at the same time, that morality is under the complete sovereignty of the human being and is subject to the categorical imperative: 'Act according to the principle which you would like to become the universal law', which applies universally to all human beings. Knowledge of the moral law is not derivable from knowledge of the physical world; it is peculiar to our understanding of our place in the universe and of our 'duty'.

Both Rousseau and Kant created systems that worked against the homogenization of the world carried on under the auspices of empiricism. They both stood up unflinchingly to the challenge of empiricism; theirs were imaginative and constructive systems, not merely defensive responses. The situation changed in the nineteenth century. Then protest against spreading materialism and positivism was almost invariably expressed from defensive positions – often from the position of despair, as in Nietzsche and some late nineteenth-century poets.

The empiricist tradition, and the entire world-view it has brought with itself, was not something inevitable and inexorable. It was one particular intellectual 'strain', which prevailed over other traditions. Those other traditions are still alive; in particular, the conviction of the unity of knowledge and values was maintained in the eighteenth, nineteenth and twentieth centuries (particularly among poets). In protesting against the pernicious pitfalls of empiricism and its offshoots, such as logical positivism, we are not wolves howling in the wilderness, but heirs of a long and great intellectual tradition.

THE ECLIPSE OF VALUES IN THE NINETEENTH CENTURY

Although the advances of natural science in the seventeenth century were enormous, traditional values still prevailed. Newton himself wrote the *Philosophiae Naturalis Principia Mathematica* to attest to the greatness, glory and perfection of God. Admittedly, empiricists Locke and Hume were already at work postulating the separation of knowledge from values.

The eighteenth century brought about the transition. The slogans of the French enlightenment were both liberating (from the tethers of the antiquated religious world-view) and at the same time ominously constraining for they paved the way to vulgar materialism, shallow positivism and the annihilation of values in the nineteenth century.

The nineteenth century marks the triumph of science and technology and an unprecedented expansion of the scientific world-view. The aggressive assertion of positivism and materialism, of which Marxism was a part; of scientific rationality and technological efficiency; of the age of industrialization, which, alas, turned out to be the age of environmental devastation, all pointed towards a brave new world in which traditional (intrinsic) values were consigned to limbo. We need to examine this process more closely in order to understand why the triumphs of science had to signify an eclipse of values.

Science did not develop in a social vacuum but as part of the unfolding new culture. The battle against petrified aspects of institutionalized religion was waged in the seventeenth and eighteenth centuries with almost the same intensity as in the nineteenth century, which was more aggressive and successful in containing the influence of religion in the realm of thought than was true of the previous two centuries. The secular, rational, science-based world-view took its place firmly on the stage. The rest seemed merely a matter of implementation. The time appeared to be near when paradise on earth would prevail.

The battle between science and religion was by no means confined to purely intellectual matters, to ways of interpreting the world around us. It was also an ideological battle; and it was an eschatological battle, for what was at stake were the 'ends' of human life. Religion represented the status quo, it was turned inward, it urged us to perfect ourselves, and to seek the ultimate reward in the afterlife. Science represented a continuous process of change; it was turned outward, and it promised salvation here on earth. In this battle religion was often in an alliance with intrinsic values, supported them and was supported by them. On the other hand, science was in an alliance with progress. The corollaries of the two opposing forces of religion and science – intrinsic values on the one hand, and progress on the other – were themselves construed as adversaries. Indeed 'progressive' and 'revolutionary' individuals in the nineteenth century debunked with equal vehemence both traditional religion and traditional values, which they somehow identified with the feudal and bourgeois ethos, regarding them as unworthy of the new epoch, in which toughness, rationality and a no-nonsense pragmatic attitude were called for.

In this climate intrinsic values were increasingly dismissed as vestiges of an obsolete world. It is therefore no wonder that new doctrines concerning values attempted, implicitly or explicitly, to serve the scientific world-view and to justify its supremacy. The doctrine of utilitarianism proclaimed that the corner-stone of our ethics and our actions should be the principle of the greatest good for the greatest number. Formulated in this way, utilitarianism does not seem to signify the submission of ethics to the dictates of science. However, the principle was soon vulgarized to mean: the greatest quantity of material goods for the largest possible number of people. This is indeed the underlying ethos of the technological or consumer society. Thus we can see that utilitarianism has become an adjunct to material progress, its ethical justification; material progress itself is an essential part of the scientific-technological world-view. A scrupulous historian might object that this interpretation does violence to the original meaning of utilitarianism, as expounded by Jeremy Bentham and John Stuart Mill, but ethical doctrines are judged by what becomes of them in practice. The ease with which utilitarianism was 'instrumentalized' and integrated into the technological society only shows how much it was attuned to the increasingly homogenized 'brave new world'. After all, Bentham and Mill were nineteenth-century empiricists of the first order. Their views embodied all the typically empiricist limitations.

Nihilism and scientism, on the other hand, overtly preached the gospel of science, enshrined facts as deities, and condemned all products of the human spirit as 'meaningless' or reactionary. One of the most striking expressions of this new tough-mindedness is Sergei Bazarov, as drawn by Turgenev in his novel *Fathers and Sons*. Bazarov, as we remember, is a robust, exuberant and enthusiastic believer in science, in materialism, and in the world in which facts and positive knowledge are supreme values. He has no use for art, for poetry, for other 'romantic rubbish'. Bazarov announces:

> 'We have decided merely to deny everything.'
>
> 'And this you call nihilism?'
>
> 'That we call nihilism.'
>
> 'Like those artists,' said Bazarov, 'I consider Raphael to be worth not a copper groat. And for the artists themselves, I appreciate them at about a similar sum.'

Bazarov is a comprehensive embodiment of the prevailing nihilism, materialism, scientism and positivism which, in their respective ways, regarded intrinsic values as secondary, insignificant, or even non-existent in the world of cold facts, clinical objectivity and scientific reason.

Now, it takes only a moment's reflection to realize that Bazarov's philosophy has won the day, that big corporations are an incarnation of this philosophy. *Bazarovism*, if I may coin the term, has become the dominant, if only implicit philosophy of the technological society – East and West. It requires one sober look to be aware that the Soviet Union is as much dominated by the Bazarovs as is this society. The mania of continuous economic growth (mistakenly identified with progress), the enshrined mode of thinking called cost-benefit analysis (mistakenly identified as the most valid methodology), and strenuous attempts to operationalize all aspects of human existence (mistakenly called the 'rationalization' of life), are all part and parcel of the same philosophy.

We are training Bazarovs in our academic institutions. Indeed, these institutions are set up to train and produce Bazarovs. The problem is severe, for even if we are dimly aware of the fact, we cannot help it: *Bazarovism, as an overall social philosophy, has pervaded the fabric of Western society and the structure of academia.*

A most alarming aspect of the situation is that the Bazarovs still consider themselves 'the torch of progress', 'the vanguard of humanity', 'the remakers of the world for the benefit of all', while in fact they serve the most crass interests of the status quo, are in the vanguard of ecological and human devastation, and embody nothing but conformity and servitude. Within a mere hundred years 'revolutionaries' and 'progressivists' have become staunch defenders of the status quo. Such a dialectic of history may startle even well-seasoned dialecticians. Over the last decade the real revolutionaries, who have attempted to rekindle our interest in the well-being of humanity as a whole, have been not the 'tough-minded' rationalists, the ones who have been 'sweeping aside the rubble of history' to pave new ways, but the 'soft-minded' believers in intrinsic values, sometimes mystically inclined, often hostile to science and progress. As a result of these distressing shifts in the meaning of the terms 'reason', 'unreason', 'liberation', and 'oppression', liberals do not know what to believe in. They invested too much in reason and progress, which were meant to provide safeguards against oppression and exploitation, but in the meantime reason has become a form of oppression and progress a force of mutilation. Herbert Marcuse has convincingly made the case for this reversal in *One Dimensional Man* and his other writings, so we need not belabour the issue here.

The intellectual climate of the twentieth century – in the economically developed countries of the West, that is – has not only favoured the rise and dominance of the Bazarovs, it has also contrived to inhibit everyone else from considering values as one of the central concerns of human thought and human life. One of the great misfortunes of modern Western thought has been the linking of intrinsic values with institutionalized religion. The bankruptcy of one form of institutionalized religion was tantamount, in the eyes of many, to the bankruptcy of religion as such, and of the intrinsic values woven into that religion. This identification was based on faulty logic. Religion, and especially intrinsic values, are not tools of the clergy to keep the masses in order (though occasionally they have been used for that); they are forms and structures, worked out over the millennia of human experience, through which the individual can transcend himself and thereby make the most of himself or herself as a human being, through which man's spirituality and humanity can acquire its shape and maintain its vitality, and through which we define ourselves as self-transcending beings. As such, intrinsic values outline and define the scope of our humanity.

The climate of the twentieth century has anaesthetized us to our own spiritual heritage. Twentieth-century philosophy has done little to remedy the situation. Logical positivists have been notorious in manifesting their *insensitivity* to the problem of values. Even outstanding thinkers and well balanced philosophers, such as Sir Karl Popper, who has gained a reputation for being an anti-positivist, offer us precious little. Indeed, it is incredible (if not actually embarrassing) how little Popper has to say about values and how pale even that is. The shadow of positivism has engulfed us all. The value-vacuum has been an inevitable outcome of the attrition of religion and the emergence of a secular world-view.

INFORMATION – KNOWLEDGE – WISDOM

Copernicus is often singled out as a divide separating the Middle Ages from modern times. His views on knowledge, however, are closer to Plato and Augustine than to modern empiricists, for all three regarded knowledge not as a stock of facts and information, held in the repository of memory, but as an intrinsic part of being human. All three thought of knowledge as inseparable from a person's actions and judgements, so that correct knowledge, in the Augustinian sense, is the basis of proper conduct. Even Newton, though considered by empiricists to be their greatest asset, was far from thinking that knowledge was 'mere information', irrelevant to or independent from man's other concerns. Newton explicitly attempted to show the perfection of God through the harmony of his universe, which, he contended, revealed itself through the unity of the laws of physics governing the behaviour of both terrestrial and celestial bodies.

Something happened between 1700 and 1900. We divided ourselves into halves. We separated our knowledge from our essence, from our values, from our transcendental concerns. Knowledge became isolated, put into a special container called brain. This container came to be regarded as a chest of tools: we pick up from this chest this or that tool for the task at hand. There is no longer the unity of man and his knowledge. There are only specialized tools to handle specialized tasks. At this point knowledge becomes mere information. Soon it becomes translated into 'bits' of computerized data. The whole process is depersonalized, mechanized, computerized.

The separation of facts from values, of man from his knowledge, of physical phenomena from all 'other' phenomena, resulted in the atomization of the physical world, as well as of the human world. The process of isolation, abstraction and estrangement (of one phenomenon from other phenomena), a precondition of the successful practice of modern natural science, was in fact a process of *conceptual alienation.* This in turn became human alienation: man estranged himself from both his knowledge and his values. The primary cause of contemporary alienation is a mistaken conception of the universe in which everything is separated and divided and in which the human being is likewise atomized and 'torn'.

Our present compartmentalization is unnatural. In order to restore our sanity and to recompose our divided selves we have to rethink certain basic premises. To begin with, we have to realize that *the state of one's knowledge is an important characteristic of the state of one's being.* This is a restatement of the view of knowledge held by Plato, Augustine and Copernicus. This view is still held among primitive societies, notably among certain American-Indian tribes.

The statement that our knowledge is an important aspect of our being, that as total bio-social organisms we cannot and do not act independently of our knowledge, is not an expression of nostalgia for paradise lost. It is a statement describing the human condition. How can we validate such a claim, particularly at a time when knowledge seems to be quite divorced from life? If the integration of relevant knowledge is indispensable for the coherence of one's life, it necessarily follows that to deprive people of such knowledge may be a source of confusion and incoherence in their lives. One does not have to be an astute observer to perceive that this is exactly what has happened in the contemporary period. Young people (and not only the young) are lost, confused and alienated because they do not have relevant knowledge to guide them; they do not have a compass, a sense of centre that would make sense of the world around them. Instead they are furnished with bits of information and data, with expertise that they so often find to be irrelevant.

It is a pathological situation; knowledge does not provide enlightenment but confusion; the amassing of information only furthers the process of alienation. The situation is especially pathological because never in the history of humankind has learning (and supposedly knowledge) been pursued on such a vast scale as today, and never has the estrangement of the individual from the world, and from fellow humans, been greater than today. The cause must lie, then, in the nature of the knowledge we pursue. Knowledge alienated from the human mind and human values in turn desensitizes and alienates the people who acquire that knowledge.

But let us be very careful when we say that this knowledge is 'irrelevant'. For in one sense it is very relevant: it is relevant to the economic system, which is mainly interested in the maximization of profit. It is relevant to the technological society as we know it. It is relevant to the conception of the world as a factory. The system of economic, ecological and human exploitation is not interested in knowledge, let alone wisdom. But it is vitally interested in information and expertise; it is interested in its own smooth functioning, which is based on technological efficiency. For this reason we furnish our students and ourselves with information and expertise, not with knowledge.

Let us ask ourselves a very general question at this point. Is there one underlying reason for this eclipse of values and all the other ills that follow from it? Perhaps the most succinct answer to this question was given by Max Scheler, who said: 'To conceive the world as value-free is a task which men set themselves on account of a value: the vital value of mastery and power over things.' We realize nowadays that this mastery has been an illusion, that we cannot subdue the world to our will without destroying, or at least seriously impairing ourselves. Nevertheless we maintain and perpetuate the system which was designed for this grand, but ultimately pitiful, folly.

There is another general question which should be raised, namely, the question of the relation of theory to practice. The separation of values from knowledge may be seen as an abstract philosophical matter on one level. But this separation is an indispensable part of the process of turning people into Bazarovs, in order to maintain the present consumer society and the conception of the world as a factory. Let us not complain that there is no relation between theory and practice. There is: ingenious theories have been created and developed in order to justify and maintain parasitic practices with regard to other people and nature at large. It should be emphasized that the system parasitizes people and nature equally. It is of the greatest importance that we understand the relation between the economic forces of a society and its conception of nature and of the universe, between our daily practices and the view we hold of the world. These broader outlooks, or world-views, imposed on us subtly and sometimes insidiously, justify and motivate our daily practices. And let us be clear that if we accept the scientific world-view with its underlying rationality and its extension – modern technology – we have lost from the start. For this world-view conduces to and justifies turning knowledge into information, values into economic commodities, people into experts. The perilous aspect of modern science lies in the consequences it has led to, and in the requirements and demands that it implicitly makes on people and the eco-system. It is useless to argue that it is not science that did the harm but the people who applied it. Knowledge is inseparable from people. Science has moulded people's minds quite as much as people have moulded science. The twilight of scientific reason, which we are witnessing today, is not necessarily the twilight of humanity. Scientific reason will have to wane in order to release us from its overpowering tentacles so that we can repair the strained relation between knowledge and values.

Which brings us once more to the phenomenon of knowledge as an inherent aspect of one's being. This phenomenon manifests itself not only in frustrated and alienated youth, whose knowledge does not guide them because they are filled with irrelevant bits of information, but also in the converse phenomenon: our veneration of and craving for wise people. Wise people are the ones whose knowledge matters, who are in the state of being in which knowledge matters, who are the integrated ones, in the sense that their knowledge serves them as human beings. We envy them because it is a state difficult to achieve in the contemporary world. Their wisdom is simply the integration of knowledge with values; it is a demonstration that knowledge is not a futile store of information but a vital force that sustains life at all levels of human existence; it is a resurrection of the universal property of knowledge, the unity of life and knowledge.

The reintegration of knowledge with values will have to take place not in order to make each of us a sage, but in order to assure the survival of humanity. It should be abundantly clear to us that we shall not be able to cope with the plethora of problems which the present (scientific-technological) mode of our interaction with nature and other people has originated, until we again arrive at a stage in which our knowledge matters to us as human beings. This will be a knowledge intertwined with values and at their service. This knowledge will be a re-embodiment, on a new level, of Plato's and Augustine's contention that to think correctly is the condition of behaving well; with the proviso, however, that to think correctly is not merely an abstract characteristic of the brain but an expression of a state of being; a combination of intellectual insight and moral power.

This state of being, which is still maintained in wise people, is akin to the state of grace. The term 'grace' is rather loaded. Every 'self-respecting intellectual' avoids it. But its past religious connotation should not deter us from making good use of it, for this term makes us clearly aware that to think well is not to think dexterously, ruthlessly, logically: to think well requires a special state of mind and of the entire being. This state of mind needs to be cultivated and nurtured as much as we cultivate – in long years of abstract thinking – the mind geared to 'scientific objectivity'.

We have a great deal to learn from oriental cultures, from the history of our own civilization, and from primitive societies alive today, in understanding, acquiring and maintaining this state of mind in which 'thinking well is a precondition of behaving well'. What is involved is not the acquisition of another piece of knowledge – of how 'other' societies thought and acted – which we shall merely append to our existing knowledge, but a change in the structure of our knowledge and in the structure of our mind which will lead, so we should like to hope, to the healing of the value–knowledge split and to the elimination of a great deal of our present alienation.

This fundamental change will resolve many specific problems which trouble us daily, such as: how can we know what research to pursue? how do we assess whether a piece of research is beneficial or detrimental? The answer to this question (in a simplified form) is: in order to pursue good research we have to pursue a good life; we have to think 'correctly' in the all-embracing meaning of the term 'correctly'. This kind of thinking is much more difficult than mere abstract, atomistic, analytical thinking. If it seems to some as if I were saying that one has to be in a state of grace in order to do good research, they are not far from the truth. For the present mutilation of the world around us, and of other people, is directly attributable to an attitude of mind which is *graceless*, which represents the disinherited mind, the subservient mind, and which is unworthy of creatures calling themselves human beings.

Should anyone seek to condemn this attitude, which I tentatively call 'grace', as a return to pre-scientific prejudices, obscurantism or the like, I would reply: why should a state of mind in which abstract entities called 'facts' are enshrined as deities be preferable to a state of mind in which intrinsic values are so enshrined? For this is all that a 'state of grace' implies. When we say, 'Dignity is an essential component of being human', 'Freedom is a necessary requirement of the concept of humanity', we in fact 'engrace' man. We have to change the world around us, and the frame of our minds, and the structure of our knowledge so that these expressions are not phrases empty of meaning.

CONCLUSION

Over the last three centuries we have redefined the world around ourselves and those redefinitions have resulted in a violation of the world about us and of ourselves. We have to discard a great deal of the 'wisdom' of the prophets of material progress, for this progress is leading us to doom. We need to remove many spurious dichotomies and distinctions, for they are often at the root of alienation in the present-day world. Above all, we have to restore the unity of knowledge and values; we have to realize that wisdom or 'enlightened knowledge' is the key to human *meaning*. We also have to develop a new comprehensive philosophy which will make a new sense of the world around us.

Bazarovism has pervaded the fabric of Western society. This Bazarovism is the cause of our value-vacuum and is responsible for the mindlessness with which we destroy nature and ourselves. The point is not to decry the insensitivity and ruthlessness of scientific objectivity once more, but to finally see that we are victims of the invisible cosmological strait-jacket which manipulates our thinking, subverts our values and cheapens our lives. We are back to cosmology, whose invisible hands choreograph the visible dance of our life.

4

From Arrogant Humanism to Ecological Humanism

AT THE NEXT WATERSHED

Traditional humanism has emphasized the nobility, the independence, indeed, the greatness of the human, who is cast in the Protean mould. This conception of the human went hand-in-hand with the idea of appropriating nature to our own ends and needs. Ecological humanism is based on the reverse premise. It sees the human person as simply a part of a larger scheme of things: of nature and the cosmos. We have to transcend and abolish the idea of the Protean (and Faustian) Man. The consequences of this reversal are far-reaching.

Ecological humanism is not just another fancy label for the view that we should be less wasteful of nature; it implies a fundamental reorientation of perception. In the past, ecology and humanism have trodden their respective roads and belonged to different ideologies. Ecology, as a movement, has predominantly focused on the *devastated environment*. It has striven for alternative solutions and remedies in order to restore wholesomeness to the environment. Humanism, on the other hand, has mainly focused on the *devastated human being*. It has striven for solutions and remedies (to injustices and alienation, through the reform of social and political institutions), in order to restore wholesomeness to the individual.

In their partial visions neither humanism nor ecology has sufficiently grasped that the plight of the environment and the plight of the human race both have the same cause, the ill effects of which are equally visible in the capitalist and communist worlds. Indeed, communist countries foolishly assumed in the past that because they were not capitalistic, they would not pose a threat to the natural world. It is now clear that the destruction of the environment in some communist countries (for example, Poland and Czechoslovakia) is amongst the worst in Europe.

Since Socrates, the philosophy of nature and the philosophy of man have developed along different, and sometimes antithetical routes. Ecology is a recent restatement of the philosophy of nature, while humanism (whatever its denomination), is an expression of the philosophy of man. This Western dichotomy between the philosophy of nature and the philosophy of man has been at the root of our mistaken notion that nature is 'there' to be harnessed, subdued and exploited.

Ecological humanism – being part of eco-philosophy – marks the return of the unitary view in which the philosophy of man and the philosophy of nature are aspects of each other. The conjunction of ecology and humanism is not arbitrary but the fruit of a perception of the essential unity of the natural and the human world. Ecological humanism requires a broadening of the concept of ecology to encompass the balance of the human environment; the natural world then becomes vested with the same 'value' as the human world. On the other hand, the ecological balance becomes a part of the human balance. As a result, the concepts of 'ecology' and 'humanism' simply merge into each other. Both ecology and humanism are a part of our enlarged vision of the evolving cosmos. In our times we understand *de novo* how essentially we are a part of nature.

Ecological humanism offers an authentic alternative to industrial society. It holds that:

- *The coming age is to be seen as the age of stewardship:* we are here not to govern and exploit, but to maintain and creatively transform, and to carry on the torch of evolution.
- *The world is to be conceived of as a sanctuary:* we belong to certain habitats, which are the source of our culture and our spiritual sustenance. These habitats are the places in which

we, like birds, temporarily reside; they are sanctuaries in which people, like rare birds, need to be taken care of. They are sanctuaries also in the religious sense, places in which we are awed by the world; but we are also the priests of the sanctuary, we must maintain its sanctity and increase its spirituality.

- *Knowledge is to be conceived of as an intermediary between us and the creative forces of evolution*, not as a set of ruthless tools for atomizing nature and the cosmos but as ever more subtle devices for helping us to maintain our spiritual and physical equilibrium and enabling us to attune ourselves to further creative transformations of evolution and of ourselves.

ETHICS AND COSMOLOGY CO-DEFINE EACH OTHER

In the scientific world-view, ethics and cosmology (the view of the universe) are completely separate from and have nothing to do with each other. This is particularly stressed by various schools of positivist philosophy, and it is clearly seen in the attitude called 'scientific neutrality'. But this is not how things are seen in pre-scientific world-views; here the ethics of a people and their view of the physical universe co-define each other. In those world-views, which survived the test of time and were found sustaining to the people that adopted them, there is a *coherence* between the value system or ethical code of the people and their other beliefs, so that the universe appears to be a harmonious place, supportive of human strivings. That coherence is gone from the scientific world-view. As John Donne cried: ''Tis all in pieces, all coherence gone.' If we find the universe a hostile and lonely place it is our culture that has made it so for us. More precisely, it is our overall view of the structure and content of the universe – how things are, which are important and which are not, which beliefs are 'justified' and which are not – the philosophical presuppositions that mould our culture, which in turn moulds the individual to respond positively to the cosmology by which the culture is originated and programmed:

cosmology → culture → individual

There is a feedback relationship between cosmology, culture and the individual human being. Within the scientific world-view the individual is estranged from the universe and indifferent if not hostile to nature, precisely because the scientific cosmology has made the universe a coldly inimical place for us. This is a restatement of the views expressed in chapter 1.

Eco-philosophy attempts to bring back the coherence between our value system and our view of the universe in order that each shall be an aspect of the other, as it is in traditional cultures. Eco-philosophy seeks to rescue the individual not with a superficial massage after which our ego is pacified while the rest of our being is still torn asunder but by means of a thoroughgoing reconstruction of our cosmology, which, with culture, constitutes the matrix of our well-being (or ill-being).

New forms of life are created out of the materials of old forms. New forms of culture must be built out of the spiritual heritage that has been handed down. Instant culture is phoney culture; instant spirituality is bogus spirituality. Consequently the ecological cosmology developed here and the new (ecological) imperative put forward here are derived from our past ethical and philosophical heritage. There is much in our past worth saving – much, indeed, that is superior to more recent acquisitions of ours. It is not, however, a resurrection of the old that is intended here but a fresh construction from materials in our moral and spiritual treasury.

The cosmology of the Bible

The cosmology of the Bible may be deemed antiquated today but it did provide the individual with a remarkable feeling of security and a great sense of belonging. In the Bible the *universe* is seen as God's personal creation, constantly supervised by God. It is a purposeful universe, which serves the causes of God. We know some of the purposes and designs of God; to that extent the universe is knowable. We do not know some other designs and purposes; to that extent the universe is mysterious and unknowable.

Humans are conceived of as the chosen creatures of God. We are of enormous significance as the protégés of God, for we were created 'in the image of God'. All other creatures are equally created by God; we have dominion over them but not the right to exploit and destroy them (as some scholars, such as Lynn White, would wish to maintain).

Values regulate the relationships between God, the Creator, and humans, his creatures. We should clearly realize that in this cosmology, values are binding *personal* relationships, and spell out the obligations of human to God, and of human to human; they do not connect humanity with the cosmos or nature or even with other living things. We live in a small and rather harsh universe, and God is full of wrath, so these values are mainly harsh prohibitions.

The scientific cosmology

Now let us look at the structure and consequences of the scientific cosmology. Here the *universe* is conceived of as an infinitely vast physical system, working according to physical laws. It has no purpose; it has no causes; it serves no cause. The universe is knowable. It is apprehensible by means of factual knowledge. And this is the only genuine knowledge. From which it follows that whatever is beyond this knowledge is either unknowable or (in a certain sense) non-existent. It is a curiously empty universe, really an empty space with some galaxies thrown in here and there, in which we are an accident rather than a consequence of anything.

Humankind is here conceived of as an insignificant piece of furniture in the infinitely vast physical universe. We are but lumps of physical matter, infinitesimal specks wandering through the immensity of physical space-time. Ultimately, according to some (La Mettrie, Skinner), we are machines and subject to deterministic laws; according to others (traditional humanists), we are the terminal point of ourselves, the point of departure and the point of arrival. But even here we have no justification or significance beyond ourselves.

Values are human-rooted and there is no transcendence beyond ourselves. When our knowledge assures us that there is nothing in the universe but physical matter and physical laws, and nothing to our destiny but to perish and be disintegrated into atoms, our only anchor seems to be our transient existence. The rise of individualism in our culture is an acknowledgement of the meaninglessness of everything beyond and above the individual. The ancient Greeks were much more individualistic than we have ever been, but they did not *need* a doctrine of individualism. Individualism in our culture, particularly in the Anglo-Saxon culture, has become a desperate search for substitutes for our lost centre. The excesses of individualism are an expression of the paranoia of an uprooted culture and people, of individuals reduced to physical matter. The individual, within the framework of Western individualism, is a savage god who is a law unto him or herself. There can be no genuine and sustaining ethic built on such a foundation.

The other strain of our secularized ethic is that of instrumentalism. When the universe is conceived of as a clock-like mechanism or a huge factory, the only things that matter seem to be physical objects, physical processes and physical transformations. Progress becomes simply material progress, and the aspirations and achievements of people tend to be thought of more and more in quantitative terms. Goods-orientated aspirations and values require quantitative assessment. Materialistic beatitude is geared to the quantitative scale. In the absence of any accepted set of intrinsic values, instrumental values tend to become more and more universal, and equated with the ultimate criterion of all values.

There is, then, a clear relationship between our picture of the world – seen as a huge factory, within the confines of which physical knowledge enables us to understand and manipulate its workings – and our growing attachment to instrumental values, through which we manipulate the world, other people *and* ourselves. The instrumental imperative eventually gives rise to the technological imperative, which demands that we should behave according to the modes dictated by technology's drive towards increasing efficiency. The technological imperative, to which we are increasingly subjected in highly developed industrial societies, is a remarkable triumph of the mechanistic, deterministic and objectivist *modus operandi* in the realm of human affairs. The brute *modus* of the inanimate physical world is now grafted on to the delicate tissue of human life.

The secularization of the world and the instrumentalization of values did not happen overnight. But once the process started to unfold and to accelerate, the results came with a swiftness and decisiveness that brought about a mechanization of the world unprecedented in history. It is curious, if not astonishing, that this process, although copiously written about, is so little understood. We still consider values as detached from our world-view, as a kind of private domain, almost independent of the vicissitudes of society and civilization. Witness those endless, insipid, impotent discussions about values, in which pretentious and sentimental claims are made to the effect that, if only we changed our hearts a little and became more charitable to each other, all would be well.

Such discussions are bound to be impotent if they do not realize that our hearts and hence our indifference to each other necessarily reflect the ethos of civilization and of the society which has conditioned us. They provide eloquent proof of the fact that there is little understanding that values are intimately connected with cosmology – that they mirror it, shadow it, justify it. It has been a part of the heritage of positivism, that ruling philosophy of our times, to separate values from cognitive knowledge, and thus from knowledge about the world, and thus from the world itself. If positivism is taken to be the most explicit part of the overall umbrella that stands for the scientific world-view, then this scientific view, as seen through the strategies and cognitive gambits of positivism, is vitally interested in keeping us confused about the relation of values to cosmology. Indeed, the scientific world-view is interested in producing Bazarovs, as I argued in chapter 3. Although much has been written on the shallowness and inadequacies of positivist philosophy, this philosophy is somehow immune and impudent – *because* it has pervaded the ethos of the whole materialist culture: through the huge network of the culture, positivist philosophy channels continuously its superficial and one-sided imperatives.

THREE ALTERNATIVES: KANT, MARX, SCHWEITZER

That all was not perfect in 'the best of all possible worlds' (i.e. Western civilization) was seen as early as the seventeenth century, notably by Pascal. Many alternatives have been formulated since. Let me touch on three of these, because they are important for the construction of our ecological cosmology.

Immanuel Kant is of particular importance here, representing as he does a pivotal point in the development of the scientific world-view, and also because he provided some answers to the dilemmas of human morality in the age of unfolding science which are of lasting value. Kant accepted the finality of physical laws. He thought, with others of that time, that physics revealed the ultimate laws governing the behaviour of the physical world. He tried to solve two dilemmas at once. The first was: how is it possible that in spite of the notorious unreliability of our senses the physical laws, which are based on sense data, are final and irrevocable? He concluded that knowledge of physical laws is not arrived at via the mechanism of the senses, but via something that makes it imperative that this knowledge is both intersubjective and irrevocable. This 'something' is the structure of the mind with its fixed categories, which the physical laws only reflect.

Karl Popper is right in saying that science has given rise to all the important epistemological and ontological problems of modern philosophy. But Popper's frame of reference, indeed his universe, is too narrow to enable him to notice Kant's second dilemma. In his discussion of Kant, Popper has little to say about Kant's attempt to solve this dilemma: if physical science provides ultimate knowledge and seems to embrace the whole universe, what is the place of humankind in this universe? Kant had too astute a mind not to realize that, if we grant to physical science the claim of its universality *over everything there is*, then we dwindle to utter insignificance. Kant's solution, as stated in the previous chapter, was: 'The starry heavens above you and the moral law within you.'

This was a separation of the physical universe from the moral universe. It was also a declaration of the complete sovereignty of the human, that is, *vis-à-vis* the alleged universality of physical laws. Hence Kant's moral imperative: we must treat every human being as an end in itself. *Act in such a way that you always treat humanity, whether in your own person or in the person of any other, never simply as a means, but always at the same time as an end.*[1] It was a far-reaching and bold response to all attempts to instrumentalize humanity, i.e. to turn the human being into a means to other ends.

There is a transcendental element in Kant's conception of the human race. Although separated from religion, the human is regarded as, in a sense, sacred. The reverence with which he speaks of the human race as an end in itself makes us aware that we are creatures beyond clay and even beyond the stars. This conception of the human, upheld by Kant and others after him, lingers on in our own conscious and subconscious awareness and has been a buffer zone against the growing encroachment of instrumental values.

Now, what about the relationship between Kant's ethical conceptions and his cosmological conceptions? It is Kant's transcendental idealism that binds the two together. The nature of human values is transcendental, and so is the nature of the physical world. Physics explores only the surface, the phenomena, the 'real' things, 'things in themselves'; the noumena are beyond the grasp of physics, beyond our understanding. Whether Kant created his transcendental cosmology in order to justify the transcendental nature of the human is an open question. But there is no doubt that he was aware that a completely physical universe, which is completely knowable and describable in terms of physical laws, leaves us little comfort and little meaning, too.

We may also wonder whether Kant's moral imperative, much admired on the intellectual level at least, has been ineffective in the social and human realm because it attempted to situate us in one world (transcendental reality) while society and civilization have consistently attempted to situate us in another empirical/pragmatic reality. To avoid schizophrenia (the natural consequence of being constantly torn by different conceptions of life), we gradually opted for empirical pragmatic reality, including the realm of values.

Kant's solutions to the epistemological dilemma, 'How can we acquire knowledge which is irrevocably certain through our notoriously unreliable senses?' and to the moral dilemma, 'How can we guarantee the sovereignty of man in a world governed by deterministic physical laws?' were both dazzling in their scope. But the relentless force which drove him was the conviction that the laws of classical mechanics were beyond refutation, and that they uniquely isomorphize physical reality (at least that part of it which is accessible to our comprehension). We no longer hold such a conviction for we consider *all* knowledge to be revocable and tentative.[2] Had Kant possessed our insight, or our hindsight, he would no doubt have thought up different solutions to both problems. His case is illuminating for two reasons. One is that great minds may conceive of marvellous solutions, even in the most constraining of circumstances (and the growing universality and rigidity of the physical structure of the universe was such a constraint). And quite another reason is that even the greatest minds are at the mercy of the assumptions of their times. All things considered, Kant's insistence on treating the human race as an end in itself is a salutary defence of the sovereignty of the human against the deterministic universe.

Another alternative was that of Karl Marx. Though far more

influential than Kant's, the Marxist alternative to the capitalist world-view, when examined at all closely, turns out to be lamentably shallow and inadequate.

In the second half of the nineteenth century and the first half of the twentieth century, Marxism was seen by many as the only worthy alternative to the capitalist system. Yet Marxism shares the fundamental assumptions of the post-Renaissance civilization which produced science, modern technology and modern capitalism. Marx was a complex and profound thinker. We do not really know which was dearer to Marx's heart: the liberation of the human *qua* being or the perpetuation of the Enlightenment *qua* science, for Marx was enormously impressed and influenced by the ideas of the Enlightenment. We do, however, know one thing for certain – that, taken as a whole, Marxism is a variation on the theme of secular salvation: the salvation of the human through material progress, science and technology.

Marxist cosmology is *aggressively* materialistic and carries even further the process of emptying the physical, social and human universe of spiritual and 'idealist' elements than the various traditional empiricisms. The label 'vulgar materialism' has therefore often been quite appropriate for Marxist philosophy. But even when it is not limited to its extreme form of 'vulgar materialism', the Marxist cosmology represents a woefully restricted universe. Moreover, that universe (as rendered by dialectical materialism and historical materialism) is always nervously guarded against the possibility of infection by idealism – too quickly dismissed as 'religiosity' or a 'remnant of the bourgeois ethos'. When every scrap of energy goes into defending a cosmology against possible infections, there is little scope for positive living under its inspiration.

It has been a major tragedy for Marxism that it has had to fight on two fronts simultaneously: against existing society and against all religion. For in fighting against religion, it has cut itself off from the spiritual heritage of humankind even more profoundly than the 'bourgeois philistine ethos' itself. Had Marxism embarked on both social and spiritual renovation, the history of this century might have run a different course. But perhaps this was not possible. The arrow of time was pointing in a different direction. 'Progressive' movements had to become even more secular, even more steeped in the anti-religious 'enlightenment', so that the drama of the secular civilization could unfold to its final act. *The Marxist ethic has proved both unsustaining and crippling, for it was built on the same inadequate world-view as our own.*[3]

Albert Schweitzer's alternative lacked the scope of either Kant or Marx, but he was perspicacious enough to realize that the basic problem was one of values. As he saw it, 'The ideals of true civilization had become powerless, because the idealistic attitude towards life in which they are rooted had gradually been lost to us.' He was also acutely aware that it was not enough to criticize civilization and its ills; one had to try to build something constructive. His constructive contribution was an ethics based on *reverence for life*. As he wrote of the moment when, after many struggles, his new ethical principle finally dawned on him: 'Now I had found my way to the idea in which affirmation of the world and ethics are contained side by side!'[4]

Schweitzer's ethic of reverence for life is a remarkable anticipation of the ecological ethic, which I develop in chapter 8. The principle of reverence for life is known and on occasion spoken of by the proponents of environmental ethics; but the content of it is only superficially absorbed. Listening carefully to the arguments and reasons for the environmental ethic, we so often find that the reason why we should take care of the ecological habitat is because it takes care of us. It would be counter-productive to destroy it, therefore we should preserve it. It is a principle of good management to take good care of our resources. So, in the final analysis, the ecological habitat becomes a *resource*. The environmental ethic is thus based on a calculus of optimization of our resources. It becomes an *instrumental* ethic: the ecological habitat is not of value in itself but only an instrument, a means of supporting us. For this reason some eco-thinkers have become emphatic that environmental ethics is not enough.

It is at this level of analysis that a fundamental difference appears between Schweitzer's principle of reverence for life which proclaims the intrinsic and sacred value of life itself – 'A man is ethical only when life, as such, is sacred to him' – and the seeming worship of the ecological niche, which is but a worship of our physical resources. Now, is there an irredeemable incompatibility between the two? Not necessarily – as I have shown in this book. However, many people in the ecology movement were subtly manipulated by instrumental values.

There is another reason why Schweitzer's teaching has had less influence than it deserves. He was a Protestant pastor to begin with, and never ceased to be one. The institutionalized Christian religion was perhaps more precious to his heart than his love of humanity. He considered the value of the Christian ethic to be universal, lasting and permanent, and did not see any conflict between the principle of reverence for life and the Christian ethic. The point of fundamental importance is this: he did not see that it was the other way around, that the principle of reverence of life and its consequences spell out a much more general ethic, of which the Christian ethic is a *particular* manifestation. Starting from reverence for life, the Christian ethic follows as a consequence, but the converse is not the case; the pronouncement of all life as sacred does not follow from the Christian ethic. Of the former, Schweitzer was sometimes aware: 'The ethic of the relation of man to man is not something apart by itself: it is only a particular relation which results from the universal one.' But he was too much a Christian to recognize the latter, namely that the ethic of reverence for life *supersedes* the Christian one.

There is a salutary lesson to be learned from Schweitzer. To preserve our sanctity and integrity, indeed to preserve ourselves as human beings, we must go beyond the phenomenon of man as conceived by traditional humanists. There is no basis for the sanctity of the human if man alone is our own ultimate reference.

THE PROMETHEAN HERITAGE

The relentless and heedless pursuit of material progress gives birth, in time, almost with the inevitability of sinister happenings in Greek tragedy, to the instrumental imperative and the technological imperative, both of which become the dominant moral imperatives of the technological society, with the devastating consequences that this entails. Yet we have to recognize that the pursuit of progress is not an aberration of the human condition, but an expression of it. When the ideal of progress is radically impoverished, however, it then becomes trivial progress and, ultimately, destructive progress.

The Promethean heritage is a part of our *moral* heritage. To be a moral agent is to be able to transcend the limitations of physical and biological determinants. To maintain a moral universe is constantly to engage in acts of transcendence. Progress in the real sense signifies the process of perpetual transcendence. The Promethean imperative stands for the necessity of transcendence. The Promethean imperative signifies progress in the real sense.

The Darwinian notion of evolution and the consequences following from it are not a proof against the divinity of the human; they do not 'conclusively reduce' us to the 'lower brutes' and ultimately, to unconscious and purposeless matter. On the contrary, looked at perceptively, even Darwinian evolution can be seen as a process of perpetual and increasing transcendence.

We have great difficulty in thinking of evolution as a benign process, for we have been conditioned to regard evolution within the framework of the competitive free enterprise system, in which evolution is construed as an expression of social Darwinism, and from which we are apparently to derive a moral lesson: either you suppress others, or others suppress you. Thus the notion of evolution is used as an ideological weapon: to justify and increase competitiveness and exploitation. The universe within this ideology is nothing more than an open market regulated by the entrepreneurial skills of those who are on top. Actually, the universe itself may be thought of as an entrepreneur, but of a more subtle variety than the entrepreneurs of the free enterprise system.

At one time it was important to emphasize our links with the rest of the animal kingdom. This is no longer the case. As Theodosius Dobzhansky argues:

> From Darwin's time until perhaps a quarter of a century ago, it was necessary to prove that mankind is like other biological species. This task has been successfully accomplished. Now a different, and in a sense antipodal, problem has moved to the fore. This is to establish the evolutionary uniqueness of man. In several ways, mankind is a singular, quite extraordinary product of the evolutionary process. Biological evolution has transcended itself in giving rise to man, as organic evolution did in giving rise to life.[5]

Evolution is a Promethean process, with hubris overwhelmingly present. At every juncture of this process, significant 'existential decisions' of great moral consequence were made. The transition from the amoeba to the fish was a great *moral* leap forward, for it enabled living matter to organize itself more satisfactorily on its way to explicit morality. The transition from the fish to the dexterous monkey was another moral leap forward. The transition from the principle 'an eye for an eye' to the principle 'Love thy neighbour and thine enemy' was an even greater leap forward. However, the announcement of the idea *'homo homini lupus'* (Hobbes) was a moral leap backward. So it is no coincidence at all that the secular post-Renaissance ideology gave birth to empiricism and capitalism and to principles justifying cut-throat competition and extolling the concept of 'man is a wolf to man' and ultimately an enemy of life itself. The fact is that the *modus operandi* of our world-view, as expressed through the working of its institutions, is undermining the biological versatility of life. We are still slow in grasping that.

We have to look at the whole thing from an evolutionary point of view, from a perspective much larger than the confines of the technological civilization. If we accept the context and framework of the technological society, then the behaviour of the individual according to the principle *'homo homini lupus'* is not only justifiable, but indeed inescapable. It is this very framework which is destructive of life at large as it diminishes variety and the integral unity of things, the very balance of living systems.

There is nothing wrong with the Promethean imperative. There is nothing wrong with progress. The Promethean pursuit within the context of the Greek world-view and Greek mythology was a form of progress. But the Promethean thrust within the totally mechanistic world, governed by instrumental values, is quite another form of progress. For here progress is reduced to material progress, and the Promethean imperative, without its vector of transcendence, becomes merely the instrumental imperative.

Now, either we take evolution, man and morality seriously and acknowledge that thinkers such as Teilhard de Chardin are of supreme relevance, or we only pretend to take them seriously, and through our indolence and lack of vision allow moral nihilism and relativism to prevail (let us remember that instrumentalism in the realm of values is a formidable ally of moral nihilism and total relativism). If we take evolution seriously, and ourselves as a special link in it, then it follows that we embark on a mobilization of our deeper faculties and facilitate new visions.

Transcendence is a part of our Promethean heritage and thus of our moral heritage. To consider the meaning of human life apart from acts of transcendence of the past, and not directed towards further transcendence in the future, would be to void it of its essential content. And the same is surely true of the meaning of the process of evolution at large. Now, since the human recapitulates and crowns the process of evolution, in-depth understanding of the nature of the human being is tantamount to an understanding of the fundamental structure of evolution itself. At this point Teilhard's conception of the human as an unfolding chapter of evolution, leading always to the next chapter, is not only justified but compelling. Teilhard's opus, properly read, gives us a clue as to how to pick up the thread of our Promethean heritage in the post-technological world.

Karl Popper's great achievement has been to show that in order to understand the nature of science we have to understand its growth, its dynamic and dialectical unfolding; we have to go beyond the mere products of science. We need to see evolution in the same light: in order to understand its nature we have to go beyond its molecular structure and beyond its logical reconstruction as merely a scientific theory; we have to understand its growth, its dynamic unfolding, its dialectics, its transformations. And these transformations are nothing short of magic.

THE NEW COSMOLOGY

We are now ready to assemble the various clues and fragments which our discussion has generated and which are parts of the alternative cosmology, and state the alternative moral imperative. In this section I shall be picking up the discussion on cosmology which we began in chapter 1. Specifically, I shall maintain that there is a powerful connection between cosmology and ethics.

Life continues through the stuff of which it is made. It would be foolishness to think that a novel paradigm of knowledge or of values could be created by *deus ex machina*. Even the most novel departures are rooted in old concepts and visions. Evolution goes on, new patterns *are* created. The process of transcendence continues.

Before we address ourselves to the new moral imperative, we need to clear our cosmological path. For if the scientific cosmology continues to hold us in its embrace, we shall be grounded and frustrated. We must no longer cherish the illusion that the scientific world-view is still our salvation and, given a chance, will deliver us to a harmonious, humane and satisfactory realm. To put the matter in the words of Erwin Schroedinger:

> The scientific picture of the real world around me is very deficient. It gives a lot of factual information, puts all our experience in a magnificently consistent order, but is ghostly silent about all and sundry that really matters to us. It cannot tell us a word about red and blue, bitter and sweet, physical pain and physical delight; it knows nothing of beautiful and ugly, good or bad, God and eternity.

The frontiers of physics have now moved into sub-atomic physics, quantum theory, and the recognition that, in the last analysis, observer and observed merge. All these fresh extensions of science fundamentally alter and indeed undermine the static, deterministic, mechanistic view of the universe bequeathed to us by Newton. Yet so far they have been acknowledged neither in the workings of our common sense nor, indeed, in our conception of scientific method.

We have a truly paradoxical situation, as contemporary science does not lend much support to what is popularly called the scientific-technological world-view. Yet this world-view, although it has been critically undermined by science itself, is powerfully sustained by a variety of institutions, including present schools and academia. What is it, then, that gives this world-view its legitimacy and sustaining force? Above all, our ideals of secular salvation, which have generated a profusion of socio-economic institutions, including the motor industry and our habits of conspicuous consumption.

We often adhere to this world-view because we feel that we must not betray the ideals of humanism. So much has been built and vested in this world-view (the tradition of the Enlightenment, the tradition of *Liberté, Egalité, Fraternité*) that we find it almost sacrilegious to question it seriously. We forget, at the same time, that the consequences of this world-view, in the second half of the twentieth century, increasingly violate humanism, enlightenment and liberty.

We are infinitely small as physical bodies. The universe is infinitely immense in its physical dimensions. Yet we needed this immensity as the laboratory to become what we are. Or rather, evolution needed this immensity and the particular density of matter, by and large homogeneous throughout the universe, and the burning of the stars, to carry on the process to the level of the human mind. For if the universe were not homogeneous in its density, then large parts of it would become so dense that they would already have undergone gravitational collapse. Other parts would have been so thin, John Archibald Wheeler argues,[6] that they would not be able to give birth to stars and galaxies. But this is not the end of the story. Why is the universe as large as it is? Why does it contain a number of particles which, according to our estimates, is about 10^{80}? And why has it existed for the length of time it has? The answer to all these questions is: in order to enable life to evolve. We are rehearsing some of the arguments that led to the anthropic principle; these arguments were put forth (by John Archibald Wheeler) in the mid 1970s, before the anthropic principle was enunciated.

Now, in order to have life, we have to have elements heavier than hydrogen, 'for no mechanism for life has ever been conceived that does not require elements heavier than hydrogen.'[7] In order to obtain heavier atoms from hydrogen, thermonuclear combustion must take place. But thermonuclear combustion 'requires several billion years of cooking time in the interior of a star'. However, in order to be several billion years 'old', the universe has to be, according to general relativity, several billion light-years in extent. So why the bigness and the composition of the universe as it is? 'Because we are here!' This is, incidentally, the conclusion of the astrophysicist Wheeler himself.[8] As we can see, a new picture of the universe has emerged. This new picture I call eco-cosmology. The new cosmology (I shall argue) is not God-centred (as in the Bible); is not human-centred (as in traditional humanism); is not matter-centred (as in the dying scientific-technological world-view), but is evolution-centred.

Within eco-cosmology, we receive an altogether new perspective on the universe, humankind and values.

The *universe* is here conceived of as evolving, mysterious, complex and exceedingly subtle in its operation. It is governed by physical laws in its segments of space-time, but these laws reflect only some aspects of its behaviour. The universe is partly knowable, but what unfathomable mysteries it may still hold can hardly be imagined. J. B. S. Haldane has said: 'The world may not only be queerer than we imagine, it may be queerer than we *can* imagine.' A variety of systems of knowledge can be accepted, as none uniquely expresses all the content of the universe. Life is as much a part of it and an essential characteristic, as is matter, stars and galaxies. In order to understand its most important characteristics, we have to understand its evolution. This evolution has worked out towards more and more complex and hierarchical structures, culminating in biological organisms and finally in the human being.

The universe is to be conceived of as home for the human race. We are not insignificant dust residing in one obscure corner of the universe; we are a cause, or at least a result of a most spectacular process in which all the forces of the universe have co-operated. This is at once a dazzling and a humbling prospect. For we are the custodians of the whole of evolution, and at the same time only a point on the arrow of evolution. We should feel comfortable in this universe, for we are not an anomaly, but its crowning glory. We are not lost in it, or alienated from it, for *it is us*.

The Copernican revolution did not signify the estrangement of the human from the world. *De Revolutionibus* is a hymn to the divine human in the divine cosmos. Only later, when the prophets of shallow materialism started to mould the course of our civilization, did we become 'reduced', and the Copernican revolution perverted. If we look at the matter clearly and carefully we find there is *nothing* in the entire corpus of physics, Newtonian physics especially, that prevents our viewing the universe as home for the human race, in the sense here discussed. It is rather our narrow vision and ill-conceived ideals that caused us to *read* into this physics the passing of the transcendent man.

The human race (within eco-cosmology) is regarded as of the utmost importance, not in its own right, but as a shining particle of the unfolding process of evolution. The origins of humankind may be in the cosmic dust, but during the billions of years of its transformations, evolution has produced such intricate, subtle and marvellous structures that the final product is nothing short of miraculous. In so far as we embody, maintain, and attempt to refine this exquisite organization, we are the miraculous, we are the sacred. Our sacredness is the uniqueness of our biological constitution which is endowed with such refined potentials that it can attain spirituality. Our sacredness is our conscious awareness of our spirituality and our inner compulsion to maintain it. Our sacredness is the awareness of the enormous responsibility for the outcome of evolution, the evolution which has culminated in us but which has to be carried on. Humankind is, in a sense, only a vessel, but vested with such powers and responsibilities that it is a sacred vessel.

Some proponents of deep ecology have accused me of undue anthropocentrism. The same people have maintained that the human race is a cancer on the surface of the earth. This I consider preposterous, and an insult to the glories of the evolutionary process. We have made our mistakes. Other species made their mistakes (and perished). Yet it takes a peculiar form of morbidity not to realize that life is glorious and that evolution is glorious. Thus, thou shalt not desecrate the beauty of evolution, and the beauty of the human condition.

Our uniqueness does not stem from being separated from it all nor from 'being the measure of all things in our own right', as traditional humanists have maintained, but from beholding the most precious characteristics worked out by life at large, from being the custodians of the treasury of evolution. We have lost some of the grandeur and glory attributed to us by older humanist conceptions. But we have gained something of inestimable value: we now form a unity with the rest of the cosmos, we are no longer alienated from it, we are a part of the cycle, woven into the rest; and the rest is woven into us: brute atoms and half-conscious cells have cooperated in order to bring us about.

Unity with the rest does not mean stagnation or dissolution into the primordial matter. Far from it; for there is nothing tranquil in this unity. Evolution has been a Promethean drama through and through, filled with sacrifice and hubris. Our life has happened as the result of innumerable acts of transcendence, some of which were steeped in blood and sacrifice. *We give meaning to our life while attempting to transcend it.* Such is the story of preconscious life. And such is the story of life endowed with self-consciousness. This conception of the meaning of life makes perfect sense for the pre-human forms of life and for human forms of life, be it high art or mundane day-to-day activity in which we want to get on with life, but 'a little bit better'. In each of these domains, we create meaning not by accepting the given, but by trying to transcend it. By the time the human race arrived, this universe was in the process of continuous transcendence, which we must continue, if only in order to exist.

Values (within eco-cosmology) regulate person-to-person relationships; they also regulate person-to-life relationships. Values are neither God-centred nor merely human-centred, but evolution-centred. In the ultimate sense, we might say that human values, the values to live by in the human universe, which includes much more than human beings, are to be related to and derived from the process of the unfolding cosmos, for this cosmos, as we have argued, is a co-defining component of humanity in its evolution. Alternatively, we could say that these values are to be related to and derived from the structure of unfolding evolution. One has to be careful in saying that values are evolution-centred, to be derived from the structure of evolution, for much depends on the meaning one attaches to the term 'evolution'. If evolution is conceived of as a process of blind permutations, happening in the pre-eminently physico-chemical universe, then, at worst, evolution-centred values may mean the sanctification of the brutal and merciless in the name of the survival of the fittest; and, at best, they may mean a worship of inanimate nature. However, if evolution is conceived of as a humanization and spiritualization of primordial matters, then the meaning of evolution in human terms spells out the meaning of human values. For values are those most refined aspects of human awareness, human dispositions and human aspirations which have made life extra-biological, which have made it spiritual, which have made it human.

The sanctity of values stems from our recognition, appreciation, indeed worship of those very characteristics of life, and the structures and hierarchies that support them, which have made life glowing in human terms. To live the life of a human being is to entertain sacredness and participate in sacredness, both of which, however, are given to us only potentially. One has to strive and labour, sometimes in great pain, to actualize this potential.

Thus within eco-cosmology, we recognize ourselves as a part of and an extension of the evolving cosmos. This evolution enables us to attribute sanctity to the human being. Our evolution-rooted sanctity makes us free of the tyranny of absolutes, in which earlier forms of sanctity were bound and, at the same time, allows us to transcend moral nihilism and relativism, with its arbitrariness and lawlessness. It needs to be emphasized over and again that present relativism and nihilism are not the result of the immorality of contemporary people, but mainly the consequence of a mistaken cosmology.

THE NEW IMPERATIVE

We do not wish to make a god of biology, of evolution, of nature. Our values are specifically and inherently human attributes. They must be expressed in human terms, that is, terms that have a comprehensive connotation for us on a variety of levels, which touch our minds, our hearts, and our bones. These values, indeed, embody, incorporate and crystallize the variety of forms of past evolution. Although, on the one hand, a product of a process and, on the other, a transitory stage, as we go on with our evolution to higher forms, these values must be considered as the unique set of human characteristics, humankind's existential anchor. They should summarize the various phases of evolution but also, and above all, be guidelines for concrete human behaviour: the principles that give meaning to human lives as lived in the universe that is both human and supra-human.

How should we look at these values which, on the one hand, are the summation of the sensitivity of evolution and, on the other, concrete guidelines for human behaviour in *this* world? What is our new moral imperative?

- To behave in such a way as to preserve and enhance the unfolding of evolution and all its riches.
- To behave in such a way as to preserve and enhance life, which is a necessary condition for carrying on evolution.
- To behave in such a way as to preserve and enhance the eco-system, which is a necessary condition for further enhancement of life and consciousness.
- To behave in such a way as to preserve and enhance the capacities that are the highest developed form of the evolved universe: consciousness, creativeness, compassion.
- To behave in such a way as to preserve and enhance human life, which is the vessel in which the most precious achievements of evolution are bestowed.

These five characteristics of the new imperative are only variations on the same theme. They all follow from the first formulation. This is not only inevitable, but highly desirable. A moral imperative must be general enough to provide a philosophical foundation for values. But it must be fruitful and open-ended enough to generate specific consequences and guidelines for action. For ultimately, we have to relate it to specific actions and undertakings in our day-to-day living; and also we must be able to derive from it the criteria for rejection of other sets of values which are incompatible with our own.

We have discussed some moral imperatives of the past, not in order to show that they are all spurious, but rather to demonstrate that they were groping attempts in the right direction. Indeed, the vital core of some of these imperatives can be accommodated in our new imperative. If we were to employ the categories and concepts of the older imperatives, we might express the content of the new imperative as comprising of the following:

- *The Promethean imperative:* or the necessity for transcendence.
- *The Kantian imperative:* or the celebration of evolution's highest achievements.

- *The ecological imperative:* or the necessity to preserve and enhance the living habitat around us.

From this new imperative, we must sharply distinguish:

- *The instrumental imperative*, which is the Promethean imperative emptied of its transcendence.
- *The technological imperative*, which is the instrumental imperative carried to its logical conclusion: the instrument, the machine, dictates the modes of human behaviour.

This presentation makes it immediately clear which part of our tradition we want to incorporate and why; and which part of our tradition we want to disinherit and why. I will now discuss the five imperatives in their turn, as they are related to our new imperative.

The Promethean component of the new imperative insists that the desire to improve, perfect and transcend our condition is inherently woven into the fabric of our life, is a *moral* urge with which we are endowed. We cannot understand life unfolding, human life especially, if we do not perceive that to go beyond – whatever the stage of our accomplishment – is in the very nature of life. In this sense, progress is not only justified, but inevitable. But it is progress towards an ever increasing transcendence and perfection. This progress cannot, on the level of human life, be separated from the attainment and enhancement of spirituality.

Thus, all progress is spiritual progress. But, as the Promethean story shows, the way to progress is paved with sacrifice, and sometimes ends with hubris. In so far as we sacrifice ourselves, consciously and unconsciously, we make ourselves instruments for the attainment of other goals. The story of evolution is a story of this kind of self-sacrifice. Each stage of evolution has made itself a means, an instrument towards achieving the next stage. And it is the same in human life. Sacrifice, selflessness and devotion are both natural and inevitable.

Altruism is a part of our nature, a part of the human instinct. To recognize oneself as human is to recognize one's capacity for altruism.

Societies which suppress altruism as a mode of social behaviour end up torn with strife, like our present society. Moreover, altruism is an *essential* part of the nature of evolution. Evolution would have long ago come to a halt if it were not endowed with altruism as its *modus operandi*. The basis of altruism is cooperation. Evolution without cooperation of its component parts would be null and void. All those theories of aggression, which revel in the apparently destructive nature of the human and which are purportedly based on evolution, seem to be quite oblivious to the work evolution has done through its altruism. It is not asserted here that aggression is not part of our heritage, but only that altruism has prevailed and will prevail, because it is in the nature of evolution. We could not live one single day, even in the meanest of societies, without altruistic behaviour occurring all the time.

Seen in its evolutionary splendour, human life is a self-burning torch. We make countless sacrifices because we think it is worth it. We make instruments of ourselves because we consider the cause worthwhile. The Promethean aspects of our life, even when we sacrifice ourselves and burn ourselves out, are often intensely satisfying to the individual – take the lonely explorer who dies for the sake of learning, enlightenment, humankind. These sacrifices are intensely satisfying because they are in harmony with the overall imperative: to preserve and enhance what is best in the species. The Promethean dimension of human life is contained in the process of ascription of meaning to life: 'We give meaning to our life while attempting to transcend it.'

But human life must never be turned into a means only. No cause is grand enough to require human sacrifice if, in the process, the human being does not fulfil himself as *human*. Hence the importance of the second part of our imperative, the Kantian imperative: treat the life of every human being as an end in itself. Expressed in evolutionary terms, this second component of our new imperative signifies the celebration of life at its present highest point. In its course, evolution creates such wonders ('What a piece of work is a man!') that to be truly aware of it, is to treat the human being as sacred. To live the life of a human being is to entertain sacredness and participate in sacredness. At this stage of evolution, humankind is an ultimate value. Yet we do not set it apart or treat it as a 'thing in itself', but in it we humbly acknowledge the workings of evolution. In this acknowledgement there is a tacit premise that evolution will go on, and that we can and will transcend ourselves and our present status. In this acknowledgement there is also a silent consent to allow the human race to turn itself into an instrument for the sake of the future, for the sake of the increased perfection of the species, and evolution at large. The dialectic between the means and the ends of human life is then painful and not easily reconcilable. In the ultimate analysis, we must never require a sacrifice from the individual for the sake of 'the future' if the individual does not fulfil his human destiny in the process. For, as we have argued, we may make ourselves into instruments, while at the same time fulfilling a part of our human destiny. We can be both ends and means at the same time. What is inadmissible is to turn human beings into *mere* means as has been done in totalitarian regimes, both ancient and recent. Here the Kantian aspect of our imperative speaks with its full resounding voice.

We make sense of the Kantian imperative, and incorporate it into our new imperative, by giving it a new evolutionary meaning. Kant did not have our cosmological insight, nor could he have been aware of our ecological dilemmas. In order to assure the sovereignty of the human person, he felt compelled to separate the human world from the physical world and then to invent 'things in themselves'. We can retain the sovereignty of the human individual while regarding him/her at the same time as a part of the evolving universe. Kant's imperative is clearly an aspect of our new imperative: in enshrining the human being, we are preserving and enhancing evolution's most accomplished creation. So-called 'inalienable rights' of the individual, which are sometimes tied to the Kantian imperative, are not only in harmony with our imperative, but clearly follow from it. Indeed, these 'rights' have a much more congruent and potent justification within our imperative than within the framework of individualism or any other situational ethics.

Human life cannot be nurtured, nursed and sustained unless we nurse and sustain the ecological habitat within the womb of which we all reside. Hence the importance of the third component of our new imperative – the ecological imperative. We are at one with the ecological habitat for it represents the forms of life of which we are a part. There is a significant difference, however, between the two propositions: 'We have to take care of the ecological habitat because it feeds us,' on the one hand, and on the other, 'We have to take care of the ecological habitat because it is a part of us and we are a part of it.' In the former case 'we' and 'it' are apart, and 'it' serves 'us'. An instrumental attitude is visibly at work here, which is to say we treat the eco-system as a means, or a resource. In the latter case, 'we' and 'it' are one, and we treat the eco-system as of intrinsic value.

If human life is to be treated with reverence, so is the life of the ecological habitat. The ecological habitat is of intrinsic value, a part of life in general. At this point, Schweitzer's principle of reverence for life must be reintroduced. We treat both human life and the ecological habitat with reverence – we treat them as intrinsic values because they represent very high achievements of the evolving universe. Schweitzer's imperative, needless to say, is congruent with our new imperative. Yet the same phrase 'reverence for life' belongs to two different cosmologies and therefore has two different meanings. Within the Christian cosmology, in which everything is God's personal property, the ethic based on reverence for life at large is an anomaly. Within eco-cosmology, which considers the universe as home for the human race, the principle of reverence for life follows naturally.

Now, there will always be conflicts, clashes and agonies within the compass of life, for we cannot sustain all forms of life. Within the structure of evolution, the more highly developed the organism, the greater is its complexity and its sensitivity and the more reason to treat it as more valuable and precious than others. In a nutshell, the exquisiteness of the human is more precious than the exquisiteness of the mosquito. In times of conflict, we care more for the life of a human being than the life of a mosquito.

We have always known instinctively that the life of a human being is more valuable than the life of a mosquito. Our new imperative gives a compelling reason why it should be so.

One is surprised indeed when some people (again within the deep ecology movement) maintain that each life is equally valuable. That we should cherish each life clearly follows from the principle of reverence for life. But it is simply not the case that each life is *equally valuable*. The advocates of so-called radical egalitarianism know this very well in their hearts – as they would not dream of taking a human life to save a mosquito or even a cat. Why then do they push their clearly untenable ideals? What we need are ideals that help life, not the ones that sound hip and fashionable.

We must take care of the ecological habitat, because it is an extension of our sensitivity. We are the guardians and stewards, not just 'users' of the eco-habitat in the same sense in which we are the guardians and stewards of human life and of our spiritual heritage. That is what the new imperative means in terms of our ecological habitats and the evolving new eco-ethics.

Let us now take a closer look at the instrumental and technological imperatives that have exerted a great deal of influence in our time, but which are *not* a part of our new imperative. Instrumental values, as representative of the instrumental imperative, have their legitimacy in the Promethean imperative. Indeed, they are derived from this imperative and in a sense are an aspect of it. For instrumental values represent an attempt to improve material conditions and thus indirectly the human condition. But the instrumental values of the industrial society have 'liberated' themselves from what is essential in the Promethean imperative: the element of transcendence. (By transcendence we mean the augmentation and enhancement of our spirituality.) Instrumental values are the ones that challenged the idea of hubris and themselves become the carriers of nemesis. They are derived from the Promethean imperative, but they are outside our new imperative, for in seizing on one aspect of our development, they simply forgot what this development is about. This is especially striking in the technological imperative which is a derivation of the instrumental imperative. The *raison d'être* of the technological imperative is *not* an enhancement of evolution at large, but an increase of industrial efficiency.

We have discussed the new imperative in terms of older concepts and imperatives, but it would be inappropriate to think of it as a mere summation of older codes, or even as a synthesis of them. Synthesis belongs to the realm of chemistry. You can make it if you know in advance how to combine the various more primitive elements. World-views and moral imperatives are not syntheses of this sort. They emerge out of human development as new species emerge in the course of evolution. None can be said to be a mere synthesis. They are more creations than combinations. Now, taking into account the intensity of our search for a non-relativistic ethic, and numerous attempts to rethink our cosmological predicament, it would appear that sooner or later someone would connect the two and show that ethics and cosmology co-define each other, or at any rate complement each other, and that a non-relativistic ethic for the future would have to be rooted in a cosmology in which the universe is conceived of as Home for Humankind. It has been the chief purpose of this chapter to demonstrate just that. What seems to me unquestionable at this point is the fact that *values can be derived from 'the laws of evolution'*; but not in any obvious or trivial sense.

When the first amoebas transcended their original state of biological being and went on to something more complex and more refined, that was an act of true transcendence; yet not one that can be characterized as endowed with divinity. However, in the process of evolution, matter went on refining itself, its sensitivity and potentiality, to the point where it created us, who in our strivings and in our actualization of the potential given to us, engage in acts of transcendence which can be characterized as carrying with them spirituality and divinity.

Sacredness is acquired in the course of evolution, not given to us on a silver platter by an omnipotent and benevolent God. Spirituality, sacredness and divinity are singular attributes of the human chapter of evolution, and these attributes have been won through many tortuous and sometimes unbelievable biological, cognitive and spiritual battles – for our original biochemical equipment did not particularly favour the development of spirituality in us.

We are fragments of emerging divinity. Spirituality and divinity appear at the end of the process of spiritualization of matter, not at the beginning. We are actualizing God, and we are bringing Him to being, so to speak, by actualizing the sensitivity–sacredness–divinity latent in us. God is at the end of the road. We are its awkward, dim, unpolished fragments for the time being.

Transcendence without God a priori given is possible, for transcendence stands for an ever-increasing perfection of our capacities and attainments. Indeed, only this concept of transcendence is justified within a truly evolutionary perspective: transcendence that is void of an original God. If we assume God at the beginning, then transcendence stands for a process of curious retardation – of going back – and not for the process of going beyond, and beyond, and beyond until we reach pure spirituality.

Eco-cosmology maintains that *we* are the universe in the making. We strive for meaning through our own existential efforts. We give meaning to the universe through our acquired humanity. We evolve aesthetic sensitivity as a part of the evolutionary process. We acquire Mind and its various cognitive capacities through our (and evolution's) strivings. We acquire spirituality as the result of our evolutionary unfolding. We acquire godliness by making gods of ourselves at the end of our evolutionary journey.

CONCLUSION

Is the programme of eco-philosophy so ambitious as to be unrealistic? The 'realism' of our present thinking is hopelessly *unrealistic*. Eco-philosophy cannot be proved by argument but must be incorporated in the structure of our lives. I believe it has already been incorporated in various places, sometimes consciously, usually subconsciously, into existing ways of life. So in a sense I have merely attempted to grasp and codify the emerging new shapes of life. Above all, I have attempted to demonstrate that eco-philosophy is a coherent philosophy, that it does not defy reason, for it is itself an expression of reason, of reason seen in its evolutionary unfolding.

Immanuel Kant asked: 'What is man?' His intention was not to describe human nature as it is, as it can be found by empirical surveys, but rather to discover the full scope of human potential.

To fulfil human potential is to transcend our present condition, and to fulfil the requirements of evolution. Our immediate and long-term biological and environmental survival depends entirely on our capacity to remake the world from within. Ecological values are nowadays an expression of historical necessity. To reach beyond is the evolutionary imperative, and it is the imperative of our present condition. Moreover, we have to reach beyond in order not to be swept away from where we are. *The transcendent and the urgent are one*. As Browning said:

> Ah, but a man's reach should exceed his grasp,
> Or what's a heaven for?

In the chapters that follow, I shall attempt to articulate further the insights of eco-cosmology and eco-philosophy. Articulation is part of creation. We must not only conceive new thoughts; we must be able to translate them into new life forms; and then we must live these new forms of life – thereby giving a palpable evidence of the viability of our new logos. Only then is the cycle completed: logos and life embrace each other and dissolve each in the other. Philosophy then becomes a form of life.

5

The Ecological Person

Woven out of the threads of the Greek dream of the power of the human mind; shrunken and twisted by the medieval matrix of the human subjected to religion; half-liberated in the period of the Renaissance; locked up into a new harness of slavery called the Industrial Revolution; intoxicated and blinded by the materialist utopia in the first part of the twentieth century; at the end of it emerges a new being – the *ecological person*.

From the valley of the Tigris, from Mesopotamia, Assyria and Babylon, arose civilizations that were new, great and proud. Within the Aegean Basin a further refinement of the human spirit occurred. The old past in the shape of turtles and orang-utans had been left behind. The horizons of a great adventure, of great spiritual achievements, of great leaps forward, were suddenly opened to dazzle the imagination and challenge our courage. However, after the extraordinary leap of Ancient Greece we seem to have succumbed to a curious spiritual anaemia. We are, as it were, intimidated by our immense potential. Blinded by the sun we have created, or perhaps envying the turtle, we started to *crawl* again. But there is no place any more for us in the family of turtles. We are doomed to a stellar destiny, cursed by the necessity of continuous transcendence. We must face again and again our immense future – in order not to betray humanity contained within ourselves. Human destiny is a spiral-like prison of unending perfectibility.

From the tunnel of the industrial consumptive jungle we are shyly gazing at a new light. We are the richer for the experience of material progress, for the freedom of movement, for the capacity to see the entire world within the walls of our own room. We are the poorer for the vanished forests and meadows, the valleys covered by the ribbons of concrete, the cultures wiped out from the surface of the earth. We are the richer for the painful experience that tells us that we must now tread along new paths. We are awakening from the materialist slumber in order to be inspired by the conception of the world as *mysterium tremendum*, an awesome mystery. Some are still asleep wrapped in the cocoon of the materialist dream. Their half-coherent somnambulant cogitations on the fulfilments of the consumptive paradise we shall leave aside.

But we are only at the beginning. We must make a leap of imagination, from the Newtonian, deterministic conception of the world, in which we are moved, determined and conditioned by inexorable physical forces, to the conception of the world in which we are the movers, co-creating the world as we create new forms of sensitivity and new forms of consciousness. We must have the courage to look 500 million years ahead; then think what *fantastic* beings we shall become by this time. But the sceptic is not impressed. He will tell us that we live *now*, not in millions of years to come. We shall reply to the sceptic: our *now* stretches over centuries and millenia. Let us at least have the courage of looking at the next millenium; and what do we see? The emergence of the ecological person.

Before we go into a further discussion of the attributes of the ecological person, let us look at our present scene and ask ourselves why the existing, that is twentieth-century conceptions of man do not, and cannot, satisfy us.

SHORTCOMINGS OF PRESENT WESTERN PHILOSOPHIES OF MAN

Twentieth-century Western philosophy has been strong and novel in logic, semantics and philosophy of language. It has been neither strong nor novel in the philosophy of man. This is, of course, understandable, if not obvious. If the universe of your discourse is so delineated that it considers as valid only the physical and the logical, then you cannot develop (within such a universe of discourse) any worthwhile philosophy of man. For the phenomenon of the human begins where brute matter ends. The philosophy of man begins when we attempt to outline and express that which is unique in us, that which goes beyond the physical, beyond the logical and the analytical. A philosophy of man, worthy of its name, must try to express our transcendent nature, for the human phenomenon, in so far as it is unique, transcends the physical and the biological, and certainly the logical and analytical.

By claiming that we are transcendent beings, I am simply asserting that a human is a product of evolution continually transcending itself. In so far as analytical philosophy avoids any notion of transcendence, in so far as it attempts to limit itself to immanent aspects of the human, and in so far as it insists on expressing the human phenomenon in analytical and descriptive terms alone, analytical philosophy is bound to miss out the human phenomenon (namely what is unique about us), and is bound to produce a philosophy that is so stilted and constrained that it is not worth having. Therefore, it is almost axiomatic that if analytical philosophy is faithful to its tools and its methodologies, it is bound to be inadequate for the expression of the uniqueness of the human race. The main predicament of analytical philosophy in relation to philosophy of man could be expressed as follows: analytical philosophy has developed powerful conceptual tools for dealing with the physical and the logical; these tools however are totally inadequate for handling the human, the existential, the spiritual. For this reason analytical philosophy is thoroughly inadequate for coming to grips with the human phenomenon. To put it in simple terms: What does it matter to get your language right if you get your concept of humanity wrong?

As a result of the lamentable inadequacies of analytical philosophy in dealing with the human phenomenon (also in dealing with ethics and aesthetics), Marxism and existentialism have come to the fore as prominent philosophies of man during the last fifty years. In my opinion, both Marxist philosophy and existentialist philosophy, particularly as expressed by Sartre, are fundamentally inadequate because they shrink the human phenomenon and rob us of our highest attributes: our transcendent nature, our divinity, our dignity; indirectly they also deprive us of higher hopes and larger horizons; ultimately they turn out to be the philosophies of despair.

A viable and sustainable philosophy of man must be a philosophy of hope and of higher aspirations; it must be a philosophy that encourages us to make more of ourselves in every dimension, including the cultural, spiritual and transcendental. The story of humankind has been, so far, the story of continuous transcendence, the story of *becoming* into ever more versatile, ever more knowing, ever more sensitive beings.

Religions and cultures have played a vital role in making of the human a symbolic animal. By creating religions and cultures we invented challenging symbols (greater than ourselves), which compelled us to make progress. We must not maintain (naïvely) that through making and using of tools we have humanized ourselves. No. The human is a symbolic animal *par excellence*, and it is through symbols that we are able to make something of ourselves.[1] What we think, we become; if we think nothing, we become nothing; if we think of symbols greater than ourselves, we may gradually become the image of those symbols. An empty skull that thinks nothing or utters trivialities leads to a trivial conception of humankind and a trivial society. The depth or shallowness of our thinking is the depth or shallowness of our reality, including the reality of the human person. The tragedy of Marxism as a general philosophy, and of the Marxist philosophy of man especially, has been that in naïvely believing that forces of production determine everything, it has dwarfed and diminished the human phenomenon by building up the human in the image of the forces of production: mechanistic, insensitive, crude. The main predicament of Marxism in relation to the philosophy of man could be expressed as follows: what does it matter to get economic relationships right, if you get the human, cultural and spiritual relationships wrong?

I said a while ago that existentialism and Marxism are

philosophies of despair and not of hope, while it is universally held that Marxism is an optimistic philosophy, indeed utopian philosophy, outlining far-reaching aspirations for the future. I shall be brief. I find these aspirations not so far-reaching and their optimism rather shallow. For the point is that the Marxist has rather limited horizons. Even if we allow for the realization of the communist utopia, what we witness at the end is a form of consumerist society. I can hear the voice of protest rising, arguing that in a communist society greed, exploitation and ruthless oppression of one class by another will be eliminated. This may very well be true, but a form of a consumer society it will nevertheless be, as it is bound to be if it eliminates the spiritual, the religious and the transcendental. Let us remember that the great aspiration for the Communist countries is to catch up with the West in terms of production and consumption.

The concept of humanity without spiritual dimensions and without any sense of transcendence is pitifully inadequate. For this reason I find Marxist philosophy intensely pessimistic, devoid of hope. I know that filling the stomachs of the poor is important. But after you have filled the stomachs, there is still life to be lived, meaning to be fused in human lives, beauty to be experienced, larger horizons to be sought, transcendental heavens to be embraced. I am also aware of some passages in the young Marx in which he speaks of man being a farmer in the morning, poet in the afternoon and a philosopher in the evening.[2] However, little should be made of these passages as they are not in harmony with the rest of Marxist philosophy, which (if it is to be Marxist), must insist on the primacy of the material over the spiritual, on the primacy of the modes of production over the forms of consciousness, on the primacy of brute material objects over the beautiful, the sensitive, the moral. After all 'being determines consciousness'. This is the basic credo of Marxism. I have gone into some detail of the Marxist conception of the human condition not because I want to demolish it, or because I think it all useless. The Marxist concept is, to say the least, inconsistent, and as such it contains some good things. My point was rather to demonstrate that it is not a philosophy of hope, that it is not a real alternative to the greedy, rapacious, exploitive concept of humankind characteristic of the free enterprise system; that, in short, the limitations of the Marxist conception are as fundamental as those of the empiricist, capitalist conception. Both are the fruit of secularism and of materialism and both suffer the consequences of a drastically shrunken vision of humankind. Both are infatuated with material growth and the idea of progress.

I have not said much about the existentialist conception of humankind. Perhaps I need not. For this conception too is the fruit of a shrunken vision; it is the consequence of secularism and materialism in which the lonely monad called the human being is desperately lost in this vast and impersonal universe. The only meaning this monad can be sustained by is to commiserate him or herself on being miserable. I hope I am not exaggerating too much and therefore making a caricature of existentialism. In point of fact, it was existentialism that made a caricature of the human being, reducing the individual to a dwarfish hunchback suffocating in his or her own juices and perversely enjoying the spectacle.

It is fascinating to learn that at the end of his life, in the last interview given shortly before his death, Sartre denounced his philosophy of despair. He said:

> I myself have never experienced despair. Nor have I ever to the slightest degree envisaged despair as a quality which could affect me. So here it was Kierkegaard who exerted a considerable influence on me.
>
> It was the fashion. I had an idea that something was missing in my personal knowledge of myself, which prevented me from feeling despair. But if other people talked about it, one had to accept that it had an existence for them. You will notice that there is hardly any trace of this despair in my work from that moment onwards: it was a thing of the time.[3]

Very strange that a leading philosopher would say 'it was the fashion . . . if other people talked about it, one had to accept that it had an existence for them.' Sartre went on to say this about hope:

> I think that hope is built into man. Human action is transcendent: it places its goal, its realisation in the future. There is hope in the very manner of action – in the fixing of a goal to be reached.[4]

So this was Sartre at the end of his life; and perhaps throughout his life; the Sartre we know little about.

Another general point of importance is this: the Goethean heritage, and the Blakean and the Nietzschean heritage are completely lost in the existential and Marxist conceptions of humankind; and, of course, in the conception represented by analytical philosophy. The main predicament of existentialism in relation to the philosophy of man could be expressed as follows: the existentialist conception rightly seeks to define us in terms unique to ourselves, but it wrongly thinks of the human as a lonely monad, a lost creature, drifting through a meaningless universe. To put it in simple terms: what does it matter to get the uniqueness of humankind right if this uniqueness signifies a suffocating veil of despair?

For a viable philosophy of man we must consider one which accounts for the uniqueness of the human phenomenon, which is a philosophy of hope, which is useful in explaining our evolutionary ascent and which gives scope and support for our future strivings which, of necessity, are transcendental strivings. The ecological conception of the human is, in my opinion, a salutary step in this direction. A terminological remark: by the ecological person I understand the transcendental-evolutionary conception of the human. Traditionally, the evolutionary approach meant reducing the human to lower forms and the exclusion of the divinity of humankind. Traditionally, the transcendental approach meant the acceptance of the divinity of the human (derived from some absolutes), and the exclusion of the evolutionary process. The ecological person seeks and finds the divinity of humankind in the evolutionary process.

Tillich makes ontology the central domain of philosophy. For Sartre, Heidegger and other existentialists, the phenomenological investigation is of prime concern. For eco-philosophy, cosmology emerges as the central domain. From the structure of the cosmos, properly conceived, follows the moral imperative as well as the understanding of humankind's essential characteristics.

THE ECOLOGICAL PERSON

The beginnings of humankind are hidden in the cosmic dust. However, through the billions of years of its 'creativity', evolution has created structures and beings so subtle and intricate that they border on the miraculous. We are one of the embodiments of this subtlety and intricacy. This is not our doing. We are only vessels in which evolution has stored and cultivated some of its more precious assets. The slow development of consciousness has led – on the level of *homo sapiens* – to the development of self-consciousness, and still later to the structure of consciousness endowed with divinity. The divinity of humankind is one of the specific crystallizations of the evolutionary endowment.

Our uniqueness does not stem from the fact that we are separated from the rest of the cosmos and *doomed* to human solitude as the existentialists maintained; nor does it stem from the fact that we are the measure of all things as Protagoras and traditional humanists wanted; but is rooted in the fact that we are the custodians of this incredible treasure-house museum which evolution has created.

In this vision of humankind we are losing part of the glamour that traditional humanists attached to the human condition, but we are gaining something far more important. We are now in union with the cosmos, an integral part of it. We are a part of the magnificent cycle in which brute atoms and the cosmic dust, as well as the first amoebas and the distant galaxies, have all cooperated to bring about life and then to bring about the human phenomenon. This conception of the universe and of our place in it is being increasingly confirmed by astrophysics and particle physics.[5]

The ecological person can be defined as a bundle of sensitivities which are in the process of continuous refinement. Sensitivities have played the key role in the progression of evolution, and in the emergence of humankind; our ascent has come as the result of the continuous enlargement and refinement of sensitivities produced by evolution, until we arrived at aesthetic sensitivities, moral sensitivities and a sense of divinity. Thus a sense of divinity is an evolutionary acquisition. The logical acumen, that is the ability to handle logic, is one of our sensitivities, and all knowledge is brought about by our cognitive sensitivities. Intuition is another of our sensitivities, and governs our capability for acquiring and handling knowledge.

The purpose of these sensitivities is the enhancement and the enlargement of life. The greater the scope of our sensitivities, the richer the life we experience and the richer the reality around us. When the eye emerged as a new form of sensitivity, and when evolution started to *see* through its creatures, this was a tremendous step forward in the evolutionary articulation of the universe. For seeing is a form of sensitivity, as is thinking and mystical ecstasy. The human can therefore be defined as the laboratory for increasing the scope of our sensitivities. Evolution itself is a still larger laboratory, really an alchemical workshop in which new sensitivities are continually created and refined.

This conception of evolution enables us to look at the whole cosmos in a new way, to re-evaluate our values, and also to restructure our values both in relation to the cosmos at large and in relation to other human beings. (I am here developing further the ideas proposed in chapter 3.) The ecological person appreciates and accepts the fact that human values, especially those which make our lives meaningful and which are the basis of our interaction with other human beings, are neither god-given nor merely subjective and conventional, and thus subject to human whims; but that their root and meaning lies in the very nature of the evolutionary process, in the process of our *becoming*: as a species, as moral beings, as self-sensitizing agents.

Looking at evolution as the process of humanization and as the process of continuous sensitizing of matter (of transforming matter into spirit), we can now clearly see that values are articulated and explicated forms of consciousness; they are the most refined sensitivities that humankind has evolved, and subsequently codified and legislated. The moral values, that is to say, those which are officially acknowledged as such, are but explicated forms of our sensitivity as regards loyalty, altruism, reverence, love and compassion.

We are all aware that moral laws and moral codes contain only a part of our moral disposition and only a part of our moral behaviour is reflected in them. There are situations in which we behave morally (and we know it), although our behaviour is not covered by any moral code. Our knowledge that there are all kinds of moral dispositions and numerous forms of moral behaviour that are not covered by any moral commandment is the result of our *moral sensitivity*, which is contained in our inner selves, and which is one of the attributes of our being. Those who do not possess this sensitivity, who cannot distinguish between the good and the evil, are radically crippled human beings and often end up as criminals. In so far as the technological society diminishes our moral sensitivity, it is guilty of a crime.

The chief importance and, in a sense, the sacred character of moral values, derives from the fact that they summarize and crown those aspects of human existence through which human life becomes light, becomes transbiological, becomes divine. To live a full life is to participate in this light and this divinity. However, light and divinity are given to us only potentially. The actualization of this potency is a difficult and thorny process, which at times appears impossible. It was almost impossible for evolution to actualize the human race, but it did; it seems almost impossible for us to actualize the divinity latent within us, but we shall.

The ecological person is the creature of evolution: it emerges at a certain juncture of the human evolutionary process (at a rather turbulent juncture we might add), and it will disappear at another juncture when evolution (through us) will transcend itself further.

Evolution goes on endlessly; it changes its forms and its aspirations; it changes the human condition. There are no absolutes in evolution, only one perpetual self-transcending flow.

THE ECOLOGICAL PERSON AND THE CELEBRATION OF LIFE

We are supported by the whole heritage of life: life unfolding, developing, emerging into new forms. We are just one form of life. We are aware of our uniqueness, and of our superiority over other forms of life. But this superiority is not so much *our* superiority as the superiority of nature, the superiority of life itself, which is capable of evolving self-consciousness and capable of writing poetry – through us.

Life has not evolved this extraordinary variety of forms in order to be extinguished by one species intoxicated by its power, by one civilization gone topsy-turvy in its one-sided development. Life is stronger, more enduring, more cunning, more extraordinary than the cunning and extraordinariness of one of its species. I believe that life will prevail in spite of us, in spite of the death wish of our civilization.

Life will find an avenue to use us cunningly to its own ends. Sooner or later life in us will alter those destructive structures that threaten not only human societies but a larger heritage of evolution. Indirectly and cunningly, our instinct for survival or, to put it in more general terms, the genius of life to assert and perpetuate itself, will make us redesign the various social institutions that are nowadays incongruous, will make us relinquish our more parasitic practices, will make us give up many of our wants. We shall survive because life will survive, because life is stronger and more enduring than any of its species. This is not an expression of blind optimism about our future, nor is it a declaration of our nobility and our ultimate goodness, but rather an expression of optimism about the future of Life. *Life will guard its assets by protecting us even against ourselves*. There is enough creative power and genius in life to do just that. Now, if anyone should insist that this conception of life as relentless and unfolding, enduring and creative, suggests and implies that Life is God, I shall not protest too much.

We can think of life as mere chemistry. We can think of chemistry as mere physics. Consequently, we can think of life as mere mechanistic interactions of physical bodies and chemical particles. And in so doing we are being 'scientific' and clearly obeying the criteria of instrumental rationality. But will this scientific thinking touch upon life as we live it? In short, we can brutalize the meaning of human life by translating it into mere physio-chemical matrices, but we cannot escape the feeling that this is cheapening the meaning of life.

Because of its extraordinary creative capacities, bordering on the miraculous, life could be called 'divine'. To say this, however, is not to preach a return to traditional religions. Yet there are aspects of traditional religions that add something significant to our substance, and if rejected or neglected seem to produce a crippling effect on our lives. *Without worship, we shrink; if you worship nothing, you are nothing.*

There are signs of a spiritual revival in many places and in many forms. After decades of existential anguish, based on the philosophy of nothingness, where the human being is aimlessly drifting through the meaningless universe, a desperately lonely particle, we are beginning to look at the human phenomenon in a new way.

Traditional religions have articulated the structures of our need to worship ideals larger than ourselves. Those structures are often mystified and frequently distorted by practice and ritual. The distortions and mystifications should not, however, obscure from our view the fact that the primary function of religious structures is to provide a framework for ideals that are inspiring and sustaining to our life.

We invest our deities with the most illustrious attributes we desire to possess, and then through the emulation of these attributes we make something of ourselves: as human beings and as spiritual beings. Our humanity is the product of our mirroring in our lives the qualities we have vested in our deities. The symbolic transformation of reality has been no less significant in the ascent of the human race than the invention of tools and of language. The role of religion in this symbolic transformation has been second to none. Religion creatively transforms reality to make us unselfish and altruistic; it inspires us with transcendental ideals which help us to live within the human family and help to reconcile us with ourselves. This is a supreme and salutary aspect of the traditional religions.

Religion, ultimately, is an instrument in our search for identity and integrity, and in our painful struggles to attain and preserve our humanity and spirituality. Life has created an arsenal of means and devices to enhance and perpetuate itself. On the level of human consciousness and human culture, it has created art and religion as the instruments for safeguarding its highest accomplishments. Seen in a broader perspective, art and even language itself are instruments of self-articulating life.

We should not be overly impressed by the secondary aspects of religion, art or language. Specifically, we must not confine our attention to their pathological aspects but look at them as vehicles of the articulation and refinement of life at large.

When conceived as instruments of the perfectibility of humankind, worship and religion have positive functions. Infatuated with the ideal of material progress, we have forgotten the many salutary aspects of traditional religion. Modern thinkers, such as Nietzsche, Marx, Engels, the Marxists and so many humanists of various denominations in our day, have concentrated purely on the secondary and negative functions of religion. In denying religion, they have so often negated the spiritual heritage of humankind. In fighting against the last vestiges of traditional religion, Christianity in particular, they have inadvertently shrunk the meaning of our existence by reducing it to our economic activity alone. We shall discuss the problem of religion further in the last two chapters.

LIFE AS A FORM OF KNOWLEDGE

There is an immense store of knowledge in us because we are alive and because we are an exquisite repository of life. We are continually steeped in our evolutionary heritage and we use the knowledge stored in us at times so cunningly and ingeniously that we cannot understand it. That is to say, we cannot understand it through the categories of accepted knowledge; we cannot *explain* this understanding in the categories permitted by accepted knowledge. Yet deep down we do understand it. The epistemology of life has to be created. At present we only have epistemologies concerned with the explanation of the physical.

It is easier to postulate that life is knowledge and that life and knowledge are linked together than to explain it. We are ill-equipped to understand the epistemology of life in us and life around us *because* our understanding has been conditioned and determined by abstract understanding, by objective understanding, by scientific understanding. Objective understanding is a part of understanding at large, for nature has endowed us with versatile minds. One of the attributes of the mind is that it can objectivize.

Objective understanding in its proper place, as a part of a larger compassionate understanding, is not to be rejected or dismissed out of hand. Complete understanding, however, includes intuition, abstraction, reasoning, objectivity, leaps of imagination. Only when objectivity suppresses other aspects of our understanding does it become a menace. Empathy is one form of understanding. We actually have less difficulty in employing empathy as a mode of understanding than in justifying it in our systems of knowledge. We often understand through empathy, but we cannot easily explain the process when we are pressed to do so.

The laboratory of the world is contained in all of us. The entire chemistry of the cosmos is circulating in you. The chains of energy are transformed into life. How does it happen that energy is transformed into life? Two grammes of energy into one gramme of life? What is this energy that becomes life? And then becomes consciousness? Take the relationship between chemistry and consciousness – we all know that it exists. If we starve the brain of oxygen, loss of consciousness follows. But this is only a tiny facet of what is there, of what takes place. We must excavate the cognitive layers of our evolution!

We so often resort to the knowledge stored in the layers of our evolution, and on occasions we have an awareness of it too. The language of the body, the language of the skin, the language of the eyes – they all have their inner grammar. What would our life have been without those languages? When my eyes meet your eyes I know instantly who you are, even if I cannot express it, either to you or to myself. I walk through life avoiding those against whom my eyes have warned me; and casting an invisible net over those whom my eyes have approved. I have knowledge in my eyes, and I know it. When I look into your eyes, you are an open quarry to me, in which I can see all the shapes chiselled out by life.

I can understand you through my skin. I can catch the quivering of your biology through the sparks of your eyes. I can submerge myself in your being, because my being and your being have been moulded by the same evolutionary forces and share the same heritage of life. Through my skin and tissues, through my senses and my mind acting in unison, I can tune in to listen to you and myself as to the music of evolution. My body, my skin, my eyes are the tentacles through which life rolls on, through which we tune in to the music of evolution, of which we are a part. To be rational is to understand the music of the spheres. To be truly rational is to combine and reconcile the rationality of the brain with the rationality of life; here lies the wisdom that transcends mere intellectual dexterity.

Is the poetry of my utterance the denial of the meaning of my words? Are things of beauty meaningless because they are not amenable to empirical verification? On the contrary, the function of beauty is the enhancement of the aesthetic and so of the biological aspects of life. *The function of poetry is a condensed symbolic articulation of life.* In the long run, research into the epistemology of life and into empathetic modes of understanding will be carried on and will be recorded in discursive treatises. Perhaps there will come a time when we won't need discursive treatises, we won't need the crutches of logic, for our understanding will be much swifter and much more direct. At present this epistemology and this mode of understanding are cultivated and used by us individually and often surreptitiously against the explicit dicta of discursive reason.

The epistemology of life[6] signifies mapping out the territories of our implicit faculties and resources of knowledge, including the subconscious, intuitive and extrasensory, which participate in our acts of perception and comprehension and guide us through the labyrinth of actual living, of which we are aware, if only dimly (and this dim awareness will have to be articulated as part of the epistemology of life), and which indelibly and uniquely determine the *modus* of our life on various levels of being, and which also determine the character of our interactions with other modes of being. We must be aware that all this happens beyond the confines and criteria of our current empirically orientated, discursive epistemologies.

The first precondition for establishing an adequate epistemology of life is the recognition that *the life process is a knowledge process*, that the knowledge of the brain cannot be separated from the knowledge contained in our elementary cells, and that abstract knowledge is only one end of the spectrum, of which the other end is knowledge of the amoebas from which we grew and with which we remain (as with everything else on the evolutionary ladder) in the cellular-blood relationship as we breathe and beat with the same rhythm of life. All forms of knowledge that we share with other forms of life, and all prediscursive or transdiscursive forms of cognition that are stored in the layers of our being, I call *biological knowledge*.

The epistemology of life will have to replace the present epistemology, which is centred around the inanimate and the dead. We will need to create new forms of learning, new schools and academia, in which the mind can open up and link itself with the glory of life, singing and expanding.[7]

My philosophical forerunners in the twentieth century are Whitehead, Teilhard de Chardin, and Heidegger. On the other side is the analytical tradition of Russell, Carnap and Austin in which the refinement of language has been pursued at the expense of our comprehension of larger philosophical problems. It is this analytical tradition that still dominates present-day academic philosophy, and it is this tradition that eco-philosophy opposes and seeks to replace. The epistemology of life is an articulation of eco-philosophy. One of the tenets of eco-philosophy is that it is committed to life (see the mandala on p. 40). Being committed to life, it must understand life and this ultimately means that it must celebrate life, not in any outward and facile merry-making but in a deep, almost metaphysical awareness of the wonderfully complex and mysterious nature of life.

The plastic environment is one of the creations of the scholarly genius of the twentieth century, but we always escape back to nature. The natural environment is the stuff of which we are made. We are oppressed by plastic environments (if only subconsciously), because life there is extinguished – and we need life around us to feel that we are alive.

The elusive nature of life *vis-à-vis* science is beautifully described by Albert Szent Gyorgi, a Nobel Laureate:

> In my hunt for the secret of life, I started my research in histology. Unsatisfied by the information that cellular morphology could give me about life, I turned to physiology. Finding physiology too complex, I took up pharmacology. Still finding the situation too complicated, I turned to bacteriology. But bacteria were even too complex, so I descended to the molecular level, studying chemistry and physical chemistry. After twenty years work, I was led to conclude that to understand life we have to descend to the electronic level, and to the world of wave mechanics. But electrons are just electrons and have no life at all. Evidently on the way I lost life; it had run out between my fingers.

Science is today the established orthodoxy. It commands the

support of the majority of educated people, if for no other reason than because this educated majority has gone through the mill of a scientifically orientated education and has been simply conditioned, if not actually brainwashed, by the precepts of the scientific understanding. However, real progress is not made by loud, ostentatious, pushy majorities, scientific or otherwise. It is made by small and obstinate minorities. From time immemorial, when the first amoebas started to multiply themselves and give birth to more complex organisms, *the story of life has been the story of deviant minorities* which, by not conforming to the established order of things, thereby evolved new characteristics and new functions. Progress on the evolutionary scale has always been achieved by tiny minorities inching their own way to produce new mutants, and ultimately new forms of life: genetic, biological, cultural, intellectual, spiritual. The poetry of life is inexhaustible. Life is not objective. Life is devouring. Life is self-transcending. The ferocious intensity of life's beat is the only rhythm worth listening to.

How this minority, which is questioning the omnipotence and omniscience of scientific-technological progress, is going to steer us into a life of sanity and alternative modes of interaction with nature and other human beings is still an open question. Martin Heidegger, at the end of his life and in a state of complete despair, maintained that only God can save us. This was no new option, relegating our responsibility to extrahuman agencies. We do not need a miracle. We need a concerted *will* to alter our own destiny. We need to cooperate with life itself in getting over the ugly bump that our technological civilization has created, ostensibly to further progress. This bump is not the first of its kind. Similar ones have been overcome in the past. But not by mere inertia and through the indolent assumption that 'the genius of life will save us'. Our wisdom and determination are part of the genius of life.

Life will prevail because it is stronger than any of its particular manifestations. Having articulated itself into awareness and enlightenment, of which we human beings are for the time being the guardians and custodians, it will not allow itself to be pushed to its lower levels. There is more to life than organic matter and its living processes. The inventiveness of nature is a part of its *modus operandi*. The genius of nature is an intrinsic aspect of its development. No organism or form of life relinquishes achievements and capacities that it has evolved in the course of its evolution. And this is also true of life at large. It will guard its inventiveness and genius for these are its most precious assets, and inherent modalities of its existence.

The cunning of life is infinite. For cunning is one of the devices through which the inventiveness of life and its march forward can be assured. Life uses us and all its other offspring to perpetuate itself. Whether this is anthropomorphizing life is beside the point. If this approach is called metaphysics, so be it. Life is certainly beyond physics, and this is exactly what the term metaphysics means – beyond physics. If the genius of life is removed from us, what is left of us? A Skinnerian pigeon, responding to basic stimuli with automatic responses? But even the pigeon is a marvel of the inventiveness of nature, and to understand the pigeon alone is to accept the metaphysics of life.

Life and evolution are one. To understand life, we have to understand evolution. To understand evolution, we have to understand life. The knowledge of evolution is the beginning of wisdom. To submit to evolution and the flow of life is not resignation and slavery but an enlightenment and a deeper comprehension of the human condition. In our understanding of the infinite riches and capacities of life, we are only toddlers.

We shall learn how to manage the subtleties of life unfolding because life will use us for this purpose. We shall learn the epistemology of life. We shall learn the awesome responsibility of accepting ourselves as fragments of God in the process of becoming. It is a glorious contingency that out of dimness may emerge radiance. We should not bemoan the fact that we are an instrument of life perfecting itself and perpetuating itself through us, for we do not have any other choice. Above all, we must try to understand that this is in the nature of things. We simply happened to be born in this particular universe which has generated life with its particular characteristics.

Life is not perfect, and evolution is full of blunders and inauspicious beginnings. The fact that our cycle of life does not coincide with the requirements of life in the long run, the fact that we happened to be born into a civilization that is death-ridden, should not worry us too much: like fallen leaves, we shall nourish the ground that will give rise to more conducive forms of life, to more intelligent human beings, to more enhancing societies. For life will prevail. And, within it, fragments and aspects of us.

In the words of *The Upanishads*:

> Be favourable unto us, o Life, with that invisible form of thine which is in the voice, the eye, and the ear, and which lives in the mind. Go not from us.
> As a mother her child, protect us, o Life: give us glory and give us wisdom.[8]

WISDOM, TECHNOLOGY AND HUMAN DESTINY

The Golden Calf has enticed the human imagination of millennia. It has assumed (in history) various forms and manifestations. In recent times it has assumed the form of technological plenty. When the pursuit of material wants signifies the destruction of higher paths leading to a more fulfilling life, when the satisfaction of the stomach means the deprivation of the spirit, then the pursuit of material wants is not worth it.

Technology has radically trivialized the scope of human destiny and the world around us. It has impoverished us by systematically directing us to trivial goals and away from higher human ideals: compassion, love, wisdom, inner peace. When we talk about technology and human destiny we cannot avoid discussing higher human ideals. For through them human destiny is realized. Not only our inner world has been trivialized by the avalanche of superficial gadgets, but also our language has been so impoverished and distorted that we have difficulties in talking about love, compassion and wisdom within the technological frame of reference. The technological epoch will stand out in history as an example of the great dexterity of the human mind; the dexterity which was so great that it eclipsed people's vision of the larger human horizons. We are still too close to this epoch to perceive it historically; but it is merely a passage of history none the less. Distinctive and unique? Yes. But all epochs that made a trace on human evolution were distinctive and unique.

Undoubtedly, the technological epoch was a great and dazzling experiment. We simply had to see where a given avenue of development would lead us to, how far a specific set of tools (derived from quantitative science and high power technology) would help us in our quest for meaning. We have inherited a technological vision of the world, and we acted upon it with good faith and high hopes. In so far as our hopes have been frustrated and the promise of the good life not fulfilled, it is for us both rational and necessary to embark on a new voyage, to seek a new vision, to pursue another path.

Within the evolutionary perspective, the technological epoch needs to be neither applauded nor condemned. Within the scope of human destiny, the technological epoch should be viewed as an experiment, of exploring yet another avenue open to humankind so that we could see what was at the end of it.

Seen on the evolutionary scale, the technological epoch has been a particular set of variations, one particular epicycle. Indeed, it has turned out to be an epicycle more than anything else; an epicycle in the astronomical sense of the term – a curious movement of the planet which gets out of its proper orbit to return to it later.

What is the proper orbit of human destiny? It is to return to those great ideals that alone make us human, which add to our stature and through which we further evolution. Evolution is the power that is second to none. Evolution is all, it is the ultimate context. It has worked its forms and shapes, its intelligence and purpose through amoebas and fish, and elephants and tigers. And it is working through us to articulate itself further.

Ridiculous it would be, indeed, if evolution created all its glories so that the technological man could be crowned as the pinnacle of all existence. Equally unwise it would be to assume that evolution has gone so far and now has exhausted itself and can go no further. It is a part of the egocentric folly of so many societies to assume that they are the ultimate stage of history. Such indeed has been the assumption of the technological society. A true evolutionary perspective should teach us some modesty.

Evolution goes on. And with it we go on: transcending the Bronze Age, transcending the Iron Age, transcending the technological age. What, therefore, is the next stage in the articulation of the human condition?

It is to the ecological person that we must look for guidance and direction in the future. The ecological person recognizes the redeeming and necessary nature of suffering, of compassion, of love, of wisdom. The ecological person envisages the human condition to be defined by at least four components.

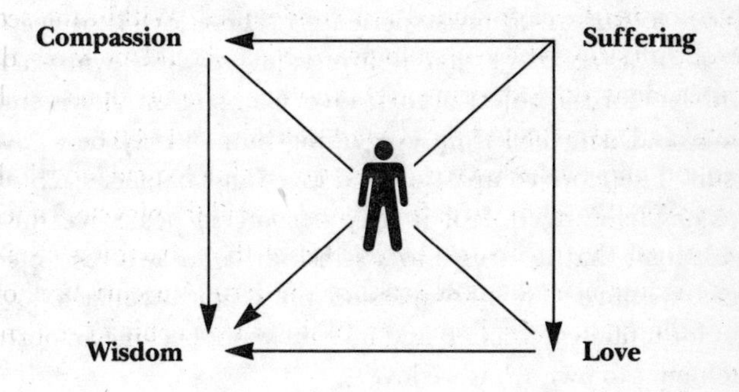

Suffering is a necessary part of both compassion and love. The person who has never suffered does not know the meaning of compassion; nor the meaning of love. Suffering, therefore, must not be avoided at all cost; as the Buddha and the Buddhists maintain, suffering must be joyously embraced as a necessary part of our existence – joyously in the sense of accepting what is ultimately beneficial to us. Spurious suffering should be eliminated; indeed it should. Whenever possible suffering should be reduced. Yet it is through suffering that we learn the sense of the human condition and the meaning of our own life.

Evolution itself has brought to being endless forms and creatures and it did so through tension and accompanying suffering. In the ideal world of pure mathematical forms it may be different, but in this world, limited and contingent, all creation and all articulation come about as the result of breaking old forms: through stress and accompanying suffering. Suffering is a mode of being in this world.

Compassion should be cultivated for it is a refined fruit of suffering; compassion is the result of a deeper understanding that all creatures suffer too, and there is nothing exceptional in our own condition. But compassion should not be made so purified and rarified that it becomes a form of complete detachment from everything. For then it makes love impossible. Love is attachment; love that is not attachment is an abstract intellectual category, not love as human beings experience it. True love is so deep and selfless an attachment that it brings joy to those who sacrifice themselves for the object of their love. Compassion that is entirely remote and detached from everything human may be a loveless pursuit. Therefore compassion and love must balance each other. Love has its dangers too; it may become so possessive through attachment, that it turns into a selfish drive to own the object or person we love. Thus compassion must not become a loveless quest for total detachment; and love must not become an obsessive attachment to own what we love.

Wisdom is the fruit of them all: of suffering, of compassion, of love. It is also the possession of right knowledge. For wisdom is knowledge enlightened by love, enshrined by compassion, refined by suffering. The right human condition is to live in wisdom, which is a form of harmony and balance, but one which must be continually re-established. Wisdom is a balance of our own being *vis-à-vis* other human beings and *vis-à-vis* the entire cosmos. We need not avoid such terms as 'cosmos' for we are creatures of the cosmos, and until and unless we find our place in it, it is unlikely that we shall find the peace within, and that kind of balance which we call wisdom.

Wisdom is not the possession of a set of permanent principles, and therefore is not to be found in *The Upanishads* or in the Bible, in the *Bhagavadgita* or in the Koran, in Blavatsky's *Secret Doctrine* or in Dante's *Divine Comedy*. Each of these great texts represents a response to specific problems and conditions. Each represents a *balance*, but for certain kinds of problems, in specific historical realities. With the flow of time, conditions, problems, and historical realities change. Wisdom is the possession of the right knowledge for a given state of the world, for given conditions of society, for given articulation of the human condition. In so far as the state of the world changes, in so far as the conditions of society change, in so far as the articulation of knowledge goes on, in so far (therefore) as the articulation of the human being proceeds, in so far as the human mind and human sensitivities become refined, we cannot embrace one structure of wisdom for all times, but *we must seek a different structure, a different form of balance for each epoch.* Wisdom is an evolutionary product; it changes its structure and manifestations with the passing of time. You cannot drink wisdom from the wells of others. Acquiring wisdom is like sculpting the inner man.

Wisdom is therefore a historical category, not a set of permanent forms but a set of dynamic structures; always to be rebuilt, restructured, readjusted, re-articulated. Evolutionary wisdom is the understanding of how the human condition changes through centuries, millennia, aeons of time. Only such a conception of wisdom can aid the race in its evolutionary voyage.

What is beyond wisdom? Enlightenment and holiness, which are beyond all knowledge; except that they are not, as they represent forms of wisdom and knowledge, which heal and illumine. Is this state reserved for the select few? Not so. Each of us possesses *rudiments* of this enlightenment and this holiness. The select ones possess them in a high degree and in a form that emanates and is visibly luminous. But even the Buddha and Jesus were not always luminous and emanating. In our own times some humble souls, such as Gandhi and Mother Teresa of Calcutta, have been luminous and emanating. The illustrious ones are the path-makers. At present they are singular exceptions; in the future they will be more or less the norm. All is natural. Enlightenment and holiness are natural. All is a part of the natural endowment of evolution, tracing out new paths through us, and through the illustrious ones.

Some would maintain that the highest attributes of the human condition, particularly enlightenment and holiness, are given to us from the outside by God or some other supernatural higher power. But supernatural powers are natural. If some find it uplifting and necessary to derive those highest attributes from trans-human deities, who are we to shatter their beliefs? But these attributes can be seen as natural ones, as attributes of evolution evolving, as the realization of our own responsibility; the responsibility which is the heightened awareness of who we are, what we are, and what we want to become.

Thus the ecological person does not seek justification for what is highest and most illustrious in us (including our divinity – as dim or as bright as it may be) in the existence and benevolence of God or some other higher power. Evolution is the answer: it is the ultimate transcending and transforming power and we are its leading edge.

This is a conception of the natural divinity of the human. The natural divinity does not take anything away from our dignity. It only makes our condition more difficult for it tells us that there is no one else to help us but we ourselves. Indeed we must help ourselves, for this is our destiny.

How does evolutionary wisdom help us to feed the poor? How does it help to overcome inflation and unemployment? How does it help us to relieve people from the stress under which they live in the technological society? First of all by making us aware that wisdom is not an agency for generating instant solutions. Instant solutions are throw-away solutions.

Evolutionary wisdom informs us that instant marriage, instant love, instant wisdom, instant society, is throw-away marriage, throw-away love, throw-away wisdom, throw-away society. Let us cultivate instead lasting and enduring society for it alone can sustain us as social beings. Let us cultivate instead lasting human relationships for they alone can fulfil us as human beings. Constant movement and change is an addiction which leads to a throw-away path.

Wisdom informs us that many chronic problems and diseases that we witness nowadays – inflation, the deterioration of our outer environment, the emptying of our inner lives – are the consequences of the pursuit of unwisdom on a phenomenal scale. Bad karma in the long run is bound to bring about bad consequences.

CONCLUSION

The ecological person directs himself to root causes and not to symptoms. The ecological person is not a miracle being, but one who perceives that lasting solutions cannot be instant solutions, that the right road may be a bit rough, a bit arduous, but this is the only road to take. The work to be done first of all is of inner reconstruction, so that we achieve some balance, some harmony within, some clarity of vision, the sense of our place in the larger universe; that we acquire, in short, some wisdom. Only then can we go on with a meaningful reconstruction of the outside world. Without wisdom we shall be like Don Quixote fighting with the windmills. The human condition has been distorted and messed up. It needs to be restored, cleansed, given a new lease of life and a new purpose.

Many will think, and indeed insist, that although the ecological conception of the human is beautiful and highly desirable, alas, it is unrealistic and impractical. Such an opinion will be an act of abdication, an expression of lack of will to wield reality into the shape of new images. All new conceptions of the human start as a dream, a vision, a utopia. They *become* a reality when men and women of purpose and determination perceive their attractiveness, and through their will and courage transform them into a reality. *All* human realities are made of webs of dreams. A web of dreams enacted continually in daily life becomes a most tangible and solid reality. The economic person is a web of dreams (of a rather inferior kind), acted upon daily. The ecological person is a web of dreams of a rather superior kind, which we *can* make a reality – if we act upon it.

We are but dust, but we are the shining dust. We are the dust that can see dust which no other dust can. Out of its shine, we create the images of God and these images transform dust into marble statues.

We are but dust, but we are the seeing dust, we are the sensitive dust. We are the dust out of which evolution makes thought. And thought transforms dust into images of perfection.

We are but dust, but we are the thinking dust. Mind is but dust, but a miraculous dust. The miraculous dust creates the universe in our own image; in the image of the shining dust that wants to become more shining.

6

Power: Myth and Reality

THE MYTH OF POWER

We are the most powerful civilization that ever existed. Yet we, as individuals, are among the least powerful people that ever existed. The roots of this paradox are part of our tragedy. By exteriorizing power, by conceiving of it as an instrument of domination of the outside world, we have deprived ourselves (as human individuals) of the power that human beings can possess. So we 'enjoy' the enormous variety of powers which we have at our fingertips, while we are powerless within. The craving for power over things (and over other people, also reduced to things) is a part of a transcendental yearning; it is an attempt to identify with a larger scheme of things. We are a tough and rational civilization, with the individual firmly entrenched as the beginning and the end of it all. Yet the transcendental yearning to go beyond one's own boundary persists in each of us. Craving for external power over things is our peculiar transcendental trip. In the words of Gregory Bateson:

> It is not so much 'power' that corrupts as the myth of 'power'. 'Power', like 'energy', 'tension' and the rest of the quasiphysical metaphors are to be distrusted, and among them 'power' is one of the most dangerous. He who covets a mythical abstraction must always be insatiable. As teachers we should not promote that myth.

Out of the many possible connotations and manifestations of power, we have chosen to enshrine one particular embodiment: power conceived as brute force for the purpose of control and domination. It is this particular manifestation of power that has become interwoven into a larger structure called Western secular civilization; and it is this form of power that causes havoc and is 'most dangerous'.

Behind the Faustian quest for power is not only the social project directed to the domination of the external environment, but something more elusive and much more profound: a new eschatology. The elevation of the myth of power to its present and dangerous position (in the Western civilization) has happened because Western peoples have given up one form of salvation and have embarked (in the post-Renaissance times) on another form. The idea of salvation was removed from heaven and placed squarely on earth. In time this salvation came to signify gratification in earthly terms alone. This meant using the earth, mastering the earth, subjugating the earth. The enjoyment of the fruit of the earth was only a part of the scheme; the other part was the enjoyment of power over the earth, over nature, over things.

Within the framework of Western secular ideology, religion became identified with stagnation and backwardness. As a result, the inward perfectibility that religion advocates and the inner powers that it wants to develop, were seen as antiquated things of the past. For this reason alone, forms of power that pertain to the spiritual realm came to be viewed as quaint and *passé*. Furthermore, secular eschatology preaching fulfilment here on earth and using the vehicle of material progress as the instrument of the physical transformation of the world became openly contemptuous of non-material human aspirations. This is the background that has led to the elevation of the myth of power.

If one is placed amidst the present merchants of power, who so intoxicate us all with their Faustian pursuits, one often thinks that power is all. If one is placed amidst learned treatises of political science nowadays – which are but a reflection of the present intoxication with power – one can hardly resist the conclusion that the pursuit of power is the dominant motif of human life. Major schools of political science, especially in the United States, lacking a historical perspective and a philosophical depth, turned themselves into indifferent describers (and *ipso facto* promoters) of power conceived as naked force. The entire realm of social science is full of words and intricate schemata but barren of real understanding. What is clearly ephemeral and pathological is so often considered normal and universal. An astute observer of the scene, the political philosopher W. Stankiewicz, concludes that as a result 'books become catalogues', not attempts at understanding but a linear rendering of the one-dimensional reality of power. 'Arguments and speculations are lacking.' It is useless to penetrate beyond the verbal façade of so-called scholarly treatises; 'They exude the staleness of newspapers and mail-order catalogues.'

I shall argue, in the course of this chapter that, ultimately, power is a property of life; it resides in living systems; thus a transfer of power is a transfer of life force. We have increasingly vested it in objects and mechanisms. This transfer – from living systems to mechanisms – was accomplished by changing the context of the whole civilization. In changing the context we have created the myth of power inspired by the demonic Doctor Faustus. Let us see how this has occurred historically.

PARACELSUS AS THE PRECURSOR OF THE MODERN QUEST FOR POWER

What are the origins of the present myth of power? There are many tributaries leading to it. Yet I think we should start with medieval alchemists who relentlessly searched for the philosopher's stone, which they envisaged as the key that would give them power to change the nature of things. It was the key to *physical* transformation that they were after, the power over things external. No one epitomizes this drive for transformation more forcefully than Philippus Aureolus Theophrastus Bombast von Hohenheim, otherwise known as Paracelsus (1493?–1541). He has emerged as an archetype of the Western mind in its restless search for effecting things external and in its quarrels with religion. He had a fascinating character: subtle, complex, profound;

he was as imaginative as he was pugnacious. His achievement, mysterious as it was, was belittled by the scholastic pedants of his time. He wrote on imagination: 'Imagination is, in itself, a complete sun – a star. If a man imagines fire, fire will result; if war, war will be the outcome . . . *Imagination takes precedence over all. Resolute imagination can accomplish all things.*' This reads more like a religious sermon worshipping the god called imagination than anything else. But he was first and foremost a physicist and an alchemist. He wrote on magic and medicine:

> Magick is a teacher of medicine far preferable to all written books. Magick alone – that can neither be conferred by the universities, nor created by the award of diplomas, but comes direct from God – is the true teacher, preceptor, and pedagogue in the art of curing the sick. I tell you that he is only a true physician who has been thus instructed, and has acquired magickal powers. And if our physicians did indeed possess these powers, all their books might be burnt and their medicines thrown into the sea – and the world would be the better for it.

There is an ambiguity here bordering on mystery. But there is also a clear declaration of trust in man's creative faculty – indeed, an expression of man's arrogance. Man is here aspiring to omnipotence, an attitude not quite in congruence with the teaching of the Church.

The fundamental question is: was Christianity already in the process of disintegration, unable to sustain the sixteenth-century people who were searching in radically new directions to find fulfilment which Christianity was no longer capable of providing? Or was it *because* of the restless mind of people at the time, bursting in all directions, that Christianity became inadequate as it could not meet the new demands of the outward-directed people? A fascinating question which we perhaps will never be able to answer.

I am told that preceding Paracelsus and the medieval alchemists were the Crusaders who clearly represented the expansive imperialist urge of the Western mind to come. True enough. But the Crusades appear to me more like the tales of adventure – a restless search for a new identity, not yet a crystallization of anticipation of a new materialist ideology. The Crusaders were, in fact, in the service of religion, if only on the surface. The secular mind possessed by the quest for power in our times represents a different phenomenon.

The early Middle Ages also saw the emergence of new technologies. But, again, we have made too much of those technologies (that is, in the twentieth century) in relation to what they were at the time. The water organ was invented around AD 1100. Watermills came into use a bit later. The former gave rise to the development of Bach's fugues filling the celestial spaces of baroque cathedrals, while the latter gave rise, in time, to the satanic mills of the nineteenth century. Which of these two developments represents a greater achievement?

In brief, the promise which was latent in medieval technology was ambiguous – by no means clear to medieval people, let alone firing their imagination. Mechanical clocks on church towers were perceived as measuring God's rhythms rather than as inventions promising a new millennium. The story of Paracelsus, on the other hand, was electrifying and became a symbol of the new strivings of Western man. In time, Paracelsus becomes Doctor Faustus. The search for the philosopher's stone gives way to the pursuit of naked power.

THE PROMETHEAN MYTH BECOMES THE FAUSTIAN BARGAIN

We are made of clay, we are made of stars, we are made of myths. Myths are the hidden springs of our thought and action. The Promethean myth is absolutely vital for the understanding of the cast of mind of Western man. Western civilization has never ceased to be Promethean in its outreach. Here lies one of the sources of the Western romance with power. This romance of late has become a sordid affair.

Prometheus is an inspiring symbol for many samaritans. He used power wisely and benevolently. There is an aura of divinity around his activities. When the use of power is not guided by wisdom and is not inspired by benevolence, then it becomes folly, and possibly a tragedy. The Greeks were Promethean through and through. Yet they were always aware of hubris, and also mindful of what Euripides said so succinctly:

> Who rightly with necessity complies
> In things divine we count him skilled and wise.

However, when the Promethean story becomes subverted by the expansive madness of the Faustian mythos, when the sense of transcendence and the sense of divinity are removed from our quest for progress, when the individual intoxication with power obscures larger paths of human destiny, and when sensuous indulgence is seen as the only salvation, then the meaning of power becomes limited to its physical, economic and manipulative aspects.

Now, we must not omit Machiavelli and his *Prince* from our analysis. *The Prince* is often seen as the direct forerunner, and indeed a forthright justification, of the concept of power as manipulation, control and domination. But the stories of the Renaissance, and of *The Prince* itself, are far more subtle and complex than the simple-minded would have us believe.

We have made of Machiavelli in political theory what we have made of Newton in physics. Newton (let us clearly bear this in mind) considered himself a theologian and considered his physics as the handmaiden of his theology. He wanted to prove that all physical phenomena in heaven and on earth obey the same set of laws. By so proving he wanted to demonstrate the harmony and perfection of God who wouldn't (in Newton's opinion) create a messy, disconnected universe with one set of laws for the heavens and another set for the earth. Physicists following him completely disregarded and ignored the theological heritage of Newton.

Likewise, present-day political science has completely ignored those subtle and transcendental aspects of both *The Prince* and the whole Renaissance tradition which, after all – particularly in its great Florentine period – was inspired by Pythagorean quests and sometimes obsessed by the exploration of esoteric mysteries. Let me quote Isaiah Berlin, who clearly shows how complex and ambiguous Machiavelli is:

> I should like to suggest that it is Machiavelli's juxtaposition of the two outlooks, the two incompatible moral worlds as it were in the minds of the readers and the collision and acute discomfort that follows, that over the years has been responsible for the desperate efforts to interpret his doctrines away to represent him as a cynical and therefore ultimately shallow defender of power politics, or as a diabolist, or as a patriot prescribing for particularly desperate situations, which seldom arise, or as a mere time server, or as an embittered political failure, or as a mere mouthpiece of truths we have always known but did not like to utter, or again as the enlightened translator of universally accepted ancient social principles into empirical terms, or as a crypto republican satirist, a descendant of Juvenal, a forerunner of Orwell, or as a cold scientist, a mere political technologist free from moral implications, or as a typical Renaissance publicist practising a now obsolete genre or in any of the numerous other roles that have been and are still being cast for him.
>
> Machiavelli may have possessed some of these attributes, but concentration on one or the other of them as constituting his essential 'true' character seems to me to stem from reluctance to face and still more discuss the uncomfortable truth that Machiavelli had unintentionally, almost casually, uncovered namely that not all ultimate values are necessarily compatible with one another, that there might be a conceptual (what used to be called 'philosophical') and not merely a material obstacle to the notion of the single ultimate solution which, if it were only realized, would establish the perfect society.

Machiavelli and the whole Renaissance tradition are made of a mould different from that of which the eighteenth-century Enlightenment, and particularly the nineteenth- and twentieth-century materialism, is made. The shift in our outlook on power did not occur in the Renaissance but comes much later as the result of a larger ideological and cosmological change, whereby Western peoples increasingly view the universe as a mechanistic aggregate to be manipulated by dexterous and powerful technology in order to gain personal security and comfort as well as the sense of aggrandizement. It is at this point that the Promethean story becomes the Faustian bargain: the dance of self-intoxication with power, whatever the consequences. The changing outlook on power represents a changing outlook on human nature.

MARX AND LENIN: MYTHOLOGIZING ECONOMIC POWER

The eighteenth century was the French century; it was the century of the Enlightenment, of the encyclopedists; it was the century in which the triumphant march of knowledge as power was increasing its momentum, in which La Mettrie published his treatise *L'Homme Machine* (1748), arguing that man is nothing but a machine.

Yet, it was also the century of J. J. Rousseau, and his *Social Contract* (1762). Rousseau's social contract was based on a *co-operative* conception of society, in which the laws were almost as highly regarded as in Plato's *Republic*, where they were accorded a sacred status. According to Rousseau, individual freedom and human dignity are safeguarded and enhanced only in so far as the individual participates in the social contract. The respect for the social contract implies the notion of democracy. Yet Rousseau regarded democracy to be beyond human reach. He wrote: 'Were there a people of gods, their government would be democratic. So perfect a government is not for men.' In spite of this contention, Rousseau was quite convinced that through the social contract we can safeguard human freedom. He was also convinced that the best social contract is one that at least approximates democracy.

The ideal of the social contract, and the view that society is a cooperative organism, undergoes a dramatic change in the nineteenth century. After the attempts of early socialists – Owen, Saint-Simon, Fourier – to bring social amelioration via piecemeal social reforms, there emerged on the stage of human history a new breed of social radicals.

Karl Marx was much inspired by the Rousseauian ethos, shared Rousseau's moral indignation and borrowed quite a bit of Rousseau's language. However, Marx was Rousseauian only up to a point. There is an important divide between the two. Rousseau's society was based on cooperation. Marx saw society as based on class struggle; ceaseless antagonism was postulated as the very essence of society. The very opening words of *The Communist Manifesto* read: 'The history of all hitherto existing society is the history of class struggle.' How very different from the way Rousseau opens his *Social Contract*: 'Man is born free, and everywhere we look we see him in chains.'

To conceive of society as a warring organism, as an entity in a ceaseless conflict, was indeed a new departure in Western thought, particularly if we consider the addition of economics as the main prism through which to view all human history. Within Marx's universe, economics became elevated to a supreme position, and at the same time placed at the centre of social conflicts. These very conflicts, we must remember, were meant to be the vehicles delivering us to social amelioration and ultimately to social justice.

The combined result of conceiving society as an essentially warring organism and of mythologizing economic power (assumed to be all-important and all-determining in human affairs) was perhaps as important in the unfolding Western secular design as the earlier mechanization of the cosmos and seeking fulfilment in material terms alone. For here, in economics deified, was a new vision which, if acted upon, would lead to far-reaching transformation. And it was acted upon; and it did lead to far-reaching transformations.

This point must be emphasized: it was a new vision, not a 'true reading of history', that Marx offered. This vision could only be validated by our willingness to act upon it, to consider it true, and thus *make* it true. The conclusion following from these arguments is quite obvious: social designs are neither true nor false in themselves. They become validated, and in a sense *made* true, if they are acted upon, accepted as true and made good by social praxis. This is exactly what has happened with the Marxian conception of society; it has been given credence and justification in the twentieth century, by being accepted as valid, by being acted upon and incorporated in social institutions.

Let us underscore two points. First, that the antagonistic conception of society is not only preached and practised in the Communist countries, but is equally in evidence in Western democracies. Paradoxically enough, while in Communist countries there is a great deal of appeal made to the altruistic nature of man, in the industrial West the conception of society as an antagonistic organism is quite unashamedly accepted and practised with ruthlessness and determination, as if the entire being of humankind was determined and nourished by economic concerns alone.

The other point that needs underscoring is that in Asian societies, India in particular (although it may be granted that the exploitation of people there is greater than in the industrial West), class warfare is less in evidence and the acceptance of the co-operative ideal of society much more pronounced. Does it mean that these societies are still 'ignorant' and have not discovered their Marx? In my opinion, it means something quite different, namely, that the Marxist vision of society as constantly at war within itself is not accepted there as a reality. Since people have not chosen to act upon the Marxist vision (as they have been nourished by other visions), it has not been validated in their social practice. To summarize: *the myths you act upon become the reality you live in*. The implication of this insight is far-reaching: all new social realities start with enacting new myths.

The Soviet Revolution has greatly helped to spread the Marxist vision of social reality. What helped the revolution was the peculiar genius of Lenin who believed in class struggle as the vehicle of change and of maintaining power. However, Lenin added a new dimension to the Marxist model: he wedded the economic concept of power to a specific organization. He was an organization man. He saw himself as an instrument of historical inevitability. He was helping history to wrest power from the capitalists in order to give it to the proletariat. The specific design of the organization to bring about the transfer of power was democratic centralism under the leadership of the Soviets. Hence Lenin's conception of communism: all power to the Soviets and – the electrification of the state.

Marrying the economic concept of power with distinct organizational structures was Lenin's contribution to the shaping of the twentieth century. Lenin, as we know, was an enthusiastic admirer of Frederick W. Taylor and of the system of Taylorism, as it was known at the time – the conveyor belt model of production. Lenin was impressed by Taylorism to the point of obsession. In emphasizing the importance of the Soviets as basic units of a larger organism, he was subconsciously trying to organize human aggregates in the conveyor belt-like transmission and multiplication of power. We witness here an ironic twist of history. Here was a man, perhaps even more despised than Marx as the red devil of communism, who was organizing his armies on a model based on the newest inventions of capitalism.

Let us now review the various components of the present paradigm of power. The Paracelsian drive for physical transformations was combined with the Machiavellian idea of ruthless manipulation, which was further combined with Bacon's idea of knowledge as power, which was further compounded by the Faustian quest for domination at whatever price, which was given a further twist by the Marxian conception of society as based on antagonism and class warfare; to which Lenin added the importance of organization, which was to some degree the product of Taylorism. We can finally identify the conceptual components of the present Western ideal of power: it is the Paracelsian–Machiavellian–

Baconian–Faustian–Marxist–Leninist–Taylorian conception of power which we cherish and which holds us in its embrace.

Each of these components of our composite idea of power came on the scene in a more or less haphazard way. The present concept of power is the result of a series of historical accidents. There wasn't any historical necessity, let alone historical inevitability, for us to inherit this particular composite which we now call *power*. Yet (strange is the human mind) we behave as if this one-sided concept of power was all there is to the meaning of power and, indeed, the only reality of power.

The prevailing Paracelsian–Baconian–Faustian–Marxist–Taylorian paradigm of power has relegated social scientists to a little corner and given them a small task: to make sense of power in the physical-economic-organizational sense, to justify the status quo. Social scientists oblige. They are also the victims of the overall conceptual myopia that affects us all. It is nothing short of astonishing that even the so-called best brains in political science (especially in the USA) have fallen victim to this myopia. Yet, not to fall victim to it entails working out a new paradigm, which is not an easy task. However, this task is now a pressing necessity; at least it is so perceived by many. I shall return to the new paradigm in the last part of this chapter.

THE EVOLUTIONARY CONCEPTION OF POWER

To understand the *nature* of power is a task much more difficult than to grasp its various manifestations. The analogy with gravity is in order here. When an apple falls from the tree we know that it could not have stopped in mid-air because of gravity. The *effects* of gravity are manifest. But what is gravity itself is another question. Gravity is not only responsible for apples falling. It also controls intergallactic spaces, including black holes, which are quite a mystery. The manifestations of gravity are numerous. To reduce gravity to the case of falling apples is a travesty of our understanding of gravity.

So it is with power. Its nature is mysterious and its manifestations manifold. To reduce the variety of phenomena that power represents to the cases of present Western institutions is a travesty of our understanding of power. The question: what is power? is perhaps as difficult as the question: what is gravity? It is much easier to describe the effects of each.

Typical of the shallow understanding of the nature of power are the endless studies on power, particularly within present political science. Even the most renowned and prestigious authors, such as Lasswell and Kaplan, are quite happy to settle for the superficial; they write: 'political science, as an empirical discipline, is the study of the shaping and sharing of power.' This kind of definition is as useful for our understanding of the nature of power, as the definition of gravity as the force that pulls apples down so that we can share them later.

In order to unearth alternative forms of power, so that we can design alternative social and political institutions, we need to understand the nature of power in some depth. We could start simply by saying that power is a relationship between object A which has power and object B which is the subject of the exercise of power. But even this preliminary formulation is faulty, for *power does not reside in objects; it resides in systems*: biological, social, cultural, spiritual. The variety of power that political science analyses resides in socio-political institutions. But these institutions comprise a small subset of *all* the relationships that are governed by the existence of power. Furthermore, the subset becomes even smaller when power is confined to economic or political institutions alone.

Power, then, is a relationship between system S_1 and system S_2. This is so even if S_1 and S_2 happen to be human individuals. For each individual is a very subtle and complex system, not just an object. Furthermore, as the result of such a relationship between S_1 and S_2, S_2 is influenced, moulded, directed or determined by S_1. All these terms – 'influenced', 'moulded', 'directed', 'determined', 'pushed around' – are easily understood in daily life. But they are not so easily comprehensible within the objectivized framework of social science. The terms 'influenced' or 'moulded' cannot be easily defined *operationally*.

When political scientists discuss the existence or the influence of power, they often behave as though it were a simple, linear and direct causation between S_1 and S_2. In actual life, we almost invariably deal with the phenomenon of multiple causation. This is especially true with regard to complex social phenomena. Given the usual complexity, there is no way of knowing whether some other factors, in addition to S_1, could have influenced S_2; and, furthermore, we also have no way of knowing whether S_2 has changed as the result of S_1 (the power exerted by S_1) or, perhaps, as the result of its own *inner* reconstruction. The scheme that social science usually assumes (when power of S_1 over S_2 is discussed) is a simplified one:

$$\mathbf{S_1} \longrightarrow \mathbf{S_2}$$

while in reality a much more complicated model is in action, namely:

$$\text{aspects of power of } S_1 \xrightarrow{\text{influence}} \text{aspects of } S_2$$

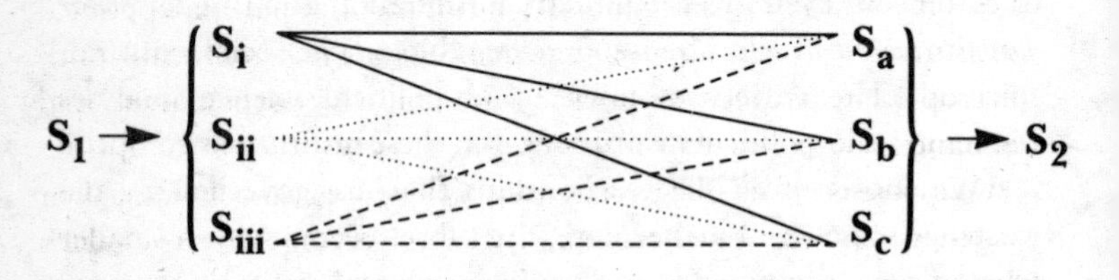

Actually, the model in real life is still more complex, as factors outside S_1 may influence (direct, determine) S_2 which are attributed to S_1. And, furthermore, there is a problem of *interpretation*. What is a cause, what is an effect, and how we determine each (if we are to use *objective* criteria) is quite a problem.

Let us be therefore aware that reality, particularly sociopolitical reality, is a very complex affair. We can, of course, reduce everything to linear models and direct causation; but this will be a triumph of a simplistic methodology, not of real understanding of the nature of power itself.

Even if we grant the claim of political science that power is predominantly the phenomenon of political institutions, social élites, and class struggle, the comprehension of the power is far from easy. It cannot be grasped by means of simplistic models, for this is a matter of multiple causation.

But power is *not* a phenomenon that can be limited to the political arena (the very phrase 'political arena' has a ring of the Roman circus of old). Power, understood as a relationship between S_1 and S_2 – on a deeper level of analysis – reveals itself as a transfer of *will*. Perhaps on a yet deeper level of analysis, the transfer of will should be understood as a transfer of life force. If the language sounds a bit mystifying, the phenomenon we attempt to describe is not.

We can now put the matter in general terms. *Power is a universal attribute of life. The source of all power is evolution.* Entropy is the inexorable march of the universe towards chaos and disorder, which means absolute death. Life is the unique organization of matter capable of combating entropy. The capacity of life to combat entropy could be called *syntropy*. Syntropy is in evidence not only in biological organisms but also in social and political institutions. A well-functioning society is a set of powerful anti-entropic structures through which life is perpetuated and enhanced.

We can now connect the concept of power with that of syntropy or anti-entropy. Power is the quality of the living universe which contributes to syntropy, or, what means the same thing, which combats entropy. This is an *evolutionary* perspective on power, and the only sensible one for understanding the nature of power in the socio-political realm and elsewhere. The ultimate purpose of all structures endowed with power is to combat entropy and enhance life. The institutions and people who contribute to syntropy represent a legitimate use of power. The institutions and people who contribute to entropy represent an illegitimate use of power. To emphasize, *the purpose of power is the propagation of syntropy through the transfer of will that is life-enhancing*. This is accomplished in the social universe through the creation and propagation of institutions that promote life.

However, such is the nature of this contingent universe that all things have a tendency to degenerate. Entropy works on us all the time. The degeneration or pathology of power reveals itself in the fact that, instead of being the instruments of syntropy, many social institutions (and people that serve them) become the instruments of entropy. From an evolutionary point of view, they represent degenerative forces, or simply anti-life propositions. That this is so with many present political institutions is obvious to everybody.

Outlined here is an evolutionary conception of power. Some might wish to discard it as it has little to do with 'present realities of power'. My response is that it has much to do with the present realities of power. The evolutionary conception of power I have outlined explains present pathologies of power; it also explains the nature of power in other realms of life – social, political, cultural, spiritual. The whole point of the first historical part of this chapter was to show that the present conception of power is but an offshoot of a myopic vision of modern Western man. Once the myopic glasses are removed from our eyes, we begin to see. The present concept of power is simply an enactment of one specific myth of power. When this myth is abolished, other manifestations and realities of power can be clearly seen. Let me now examine these other realities of power.

POWER AS AUTHORITY: SHAMANS, MEDICINE MEN, GANDHI, KHOMEINI, WALESA

The enormous shadow of the concept of power as brute force is too conspicuous to be ignored. Yet, behind this shadow there are other realities. Let me start with a most controversial one, Iran. Ayatollah Khomeini was an old man while in exile in Paris. And he did not have any power. Yet, by defying the 'reality' of power and all those who derided him as a senile man who should look after his goats and grow radishes, he rose to real power. The situation no doubt has been extraordinarily complex. But there is a single thread that runs through the whole story. The 'knowledgeable people', political scientists, expert journalists and the lot, advised the Ayatollah, first, when he was in exile, to retire quietly as he did not understand the reality of power, and in so doing they piled a great deal of abuse on him. Then, after the fall of the Shah, Khomeini was advised to be reasonable and realistic, accept a compromise and cooperate with the new government of the prime minister Bakhtiar. The arguments were the same: Khomeini was a fool as he could not accommodate himself to the realities of the world. He certainly would not. At the time he still did not have power in the worldly sense.

Then he went to Tehran, took power with his naked hands, and proceeded to defy every prediction and precept as to what power is. He was a holy fool, able to rule through the benevolence of Providence. He was perceived rather differently in the West, which constantly tried to cut him down to size and interpret him through categories 'sensible' to us. The West, however, was not competent to judge him impartially. Whoever cares to look back at the predictions of experts realizes that he made fools of all the 'knowledgeable' people who continually told him what the world and power are about. Obviously, if *they* had been half right, Khomeini would have been on the rubbish heap of history many times over.

In comparison with the real power over the destiny of his people that Khomeini possessed, the president of the United States appears just a pariah, and his power but illusory. Not only is he a victim of the Senate and the Congress; he is, above all, the victim of mass media. I am leaving aside the question of whether Khomeini's exercise of power was for better or worse. History will sort out this problem. But even here there are pitfalls, for Western history certainly is not going to be kind to him.

The case of Ayatollah Khomeini is so controversial and so many-faceted that it lends itself to almost any interpretation. Let us therefore consider a simpler one, and more telling for that. Let us look at the strike of the Polish workers at Gdańsk in the summer of 1980. It was a *par excellence* political struggle for power. Or was it? Strange as it may seem, it was a repetition of the Khomeini story – on another level and in different circumstances. At the beginning the workers did not have any power. Their strike was not even recognized by the authorities who had all the power, including the tanks that crushed the workers' protests in 1970, in the same city of Gdańsk. All the rational arguments, recognizing the 'reality' of power, were against the workers. Even the Catholic Church and Cardinal Wyszynski thought that the workers should compromise and be *reasonable*. Against all the auguries, and indeed against history, which told them that independent trade unions can never be recognized in a Communist country, the workers prevailed. Their victory was as great as it was unprecedented.

How did they do it, being so powerless to begin with, and yet ultimately so powerful? Those who are prepared to die for their cause are more likely to succeed (even without resorting to arms) than those who are sitting on the fence and always willing to compromise. This was one of the reasons. The ability to face death and the willingness to stand up and be counted, whatever the consequences, is a great moral force, and indeed a hidden source of power: political and non-political.

But a more immediate explanation of the extraordinary success of the Polish workers at Gdańsk in 1980 was the emergence of a leader, Lech Walesa, the leader who came from nowhere and did not have any power to begin with. At the end of the strike, he had the authority and power which many of the so-called 'world leaders' could only envy.

Still more surprising is the fact that when one looks at the events as they unfolded themselves, one realizes that in one sense it was not a political struggle at all. It was a religious ceremony! The whole place, where the workers were on strike, was decorated with pictures of the Madonna, of the Pope, of the saints. Beautifully arranged flowers decorated all these pictures. The atmosphere in the shipyard was both festive and reverential. Drinking vodka was strictly prohibited! This restraint from drinking was an act of votive prayer. You must know something about the drinking habits of Polish workers to appreciate this fact.

The whole situation borders on the bizarre and does not lend itself to any 'rational' explanation: to conduct a political struggle as if it were a religious ceremony is ridiculous enough; to envisage such a spectacle in a Communist country would seem impossible. However, the reality of the events has denied the rationality of our shallow explanations.

That was Poland and Walesa in the late 1970s. Then came a brutal suppression of Solidarity, which was declared non-existent. But like a phoenix it would rise from the ashes year after year, in spite of impossible odds, and in spite of all the declarations from the government that it did not exist. Then came a Nobel Peace Prize for Lech Walesa. Then another incredible event – the formation of a non-Communist government in Poland in 1989, to which Walesa contributed immeasurably. And then came a series of even more incredible uprisings – 'the Autumn of Nations', whereby East Germany, Bulgaria, Czechoslovakia and Romania abolished the Communist tyrannies to enter the era of a new freedom. In Poland the culmination came in 1991 with the free election that saw Walesa made president of his country. Walesa's case is a spectacular example showing how moral authority, in appropriate circumstances, may elevate a person from a humble electrician to a statesman being entertained by the Queen at Windsor Castle.

All these events are simply unaccountable by the past paradigm of power equated with physical force. Conventional political analysts and expert political scientists, who have been used to the yardstick of power as coercion, have great difficulty in reconciling themselves to the new European landscape and to the new concept of power that has triumphed. Only China proved to be an exception; but the massacre of June 1989, in which brute power prevailed so tragically, met with a universal revulsion, thereby demonstrating that as human beings we do not accept the existence of brute physical power as an inevitability.

We are in a dilemma, but only if we limit the notion of power to its naked physical-economic aspects. The point is that the real exercise of power has more to do with authority, particularly moral authority, than with the possession of physical means of controlling people. The notion of authority is more encompassing than the notion of power. What is authority? To answer this subtle question would require a volume. Suffice it to say that authority is a *moral* power which impels people to do things which otherwise they would be unwilling to do; it impels them to do so in the name of higher human ideals or in the name of the causes that transcend their individual interests.

This conception of authority is broader than it is usually assumed in political science. Social scientists sometimes do attempt to convey the idea of authority as a moral power. But they think that they must be neutral and give no consent to normative notions. Hence their definitions are contorted euphemisms. However, authority as a moral power is a normative, value category. Consequently its definition as 'a quality of communication by virtue of which it is accepted' is such a sad mishandling of language. For even this pedestrian, pseudo-objective definition of authority cannot escape the normative context. For what is a *quality* of communication, if not some compelling moral power? And what is this something in *virtue* of which it is accepted, if not something that goes beyond the mere idea of communication? Of the various conceptions of authority, Bertrand de Jouvenal's comes closest to mine, as he insists that: 'The phenomenon called "authority" is at once more ancient and more fundamental than the phenomenon called "state".' De Jouvenal deserves to be studied, and not only acknowledged in footnotes.

Power without authority is illusory. This is why Jesus prevailed over the Caesars. This is why Gandhi prevailed over the British. This is why Khomeini prevailed over the Shah. This is why the workers at Gdańsk prevailed over the Central Committee in Warsaw.

Earlier on, I discussed power as a relationship between S_1 and S_2 and suggested that power so conceived can be further analysed as a transfer of will, and ultimately as a transfer of life force. By accepting this interpretation of power, at least as a hypothesis, we can see that a blade of grass shooting through the surface of a concrete road is a manifestation of power so conceived. There seems to be in a blade of grass (under certain conditions) this extraordinary power which enables it to pierce through the concrete in spite of all rational arguments to the contrary.

If we take the case of Gandhi *vis-à-vis* the solidity of the British empire, the analogy with the blade of grass (making its way through the surface of the concrete) impresses itself on us irresistibly. I have tried to argue that all social and political institutions must ultimately be related to, and interwoven with, the larger network of life, of which the spiritual life of humanity is an inherent aspect. Gandhi's moral authority can be envisaged as a shoot of healthy life piercing through the concrete of a petrified life.

There are still other forms of power that have been important to past cultures: the power of shamans, of medicine men (including medical practitioners of our times); the power of holy men and of healers.

The source of power of these people seems to be of an altogether different kind from the source of power of political institutions. Yet there is a connection between political power in the modern sense (conceived as power of domination) and power as attributed to medicine men, healers and holy men. The power of holy men, shamans, healers and holy fools resides in the extraordinary personal faculties that they possess. This makes them special. Through these faculties they can accomplish things which ordinary people cannot accomplish. In traditional cultures there was a firm link between people of authority and people of those special inner powers. Authority was often the fruit of the inner power those rare individuals possessed. Authority, in its turn, was translated into political power. The chief was vested with political power *because* he had authority; and he had authority *because* he possessed those extraordinary faculties that made him a medicine man or a man of vision. Or should we say, because he allowed himself to be an agency, a wellspring of life force manifested as human power. Let us summarize.

I have distinguished three realities of power:

- Power conceived as *domination*, usually referred to as political power.
- Power conceived as *authority*, which manifests itself in a transfer of will from S_1 to S_2, whereby S_2 feels obliged to act as the result of a moral imperative.
- Power conceived as the possession of *extraordinary inner faculties*, whereby things are accomplished *as though* by magic, which only means that those who do not possess those extraordinary faculties are at the mercy of those who do possess them.

In Western culture, or should I say in *present* Western culture, dominated by the mechanistic paradigm, power as authority is belittled, and power as possession of extraordinary faculties almost completely ignored, much to our peril. Gandhi's political power was no doubt the result of his moral authority. His moral authority, in turn, seems to have been the result of his special faculties which made him a holy man (the term a 'holy fool' would be equally appropriate), whereby people were inspired and renewed by his mere presence.

The tough-minded rational pragmatist may argue at this point that what applies to the sentimental and mystically inclined Indian people does not apply to Western people. But the tough-minded pragmatist is usually constrained in his perception, and much more so in this instance. A politician endowed with power, conceived as moral authority, is so often one who possesses charisma. Charisma is another name for possessing power in the sense of being endowed with extraordinary faculties. John F. Kennedy possessed such a charisma, and he wielded power as a moral authority. For this reason he has become a legend – in spite of the machinery of the government that wanted to thwart him. Charisma and authority are irrelevant to the conception of power in mechanistic perception. When the machine defines and determines the scope and extension of political power, people may face an election in which they are deprived of choices, as in the presidential election in the USA in 1980.

Let us draw some conclusions. Power cannot be conceived in a vacuum. Its meaning is determined by the entire social context. It is the framework that determines the reality of power. To understand the meaning of power is to understand the context within which it operates. We have profoundly changed the context of Western society since medieval times, and correspondingly we had to change the concept of power. Tell me what the ideals of your society are and I will tell you what concept of power your society cherishes.

The triumph of the Western myth of power has been the triumph of *Western context* which means the triumph of Western secular ideology and its concept of material progress. This 'triumph', of course, includes industrial capitalism and industrial socialism. As we have seen, the Marx–Lenin contribution to the shaping of our notion of power has not been negligible. Communist countries joyously accepted the secular ideology of the post-Renaissance times, and, with it, other components of the Faustian conception of power.

But the triumph of the Western concept of power has by no means been universal – even at the time when entire nations and continents succumbed to the mesmerizing glamour of material progress. For, in the heart of the new leviathan, the United States, there have been enclaves of an alternative reality, where 'old shamanism' has prevailed, where the belief in power as moral authority has been upheld, and where the belief in those special inner capacities of the human has been strongly cherished. I refer to American Indians. They did not accept the context; and, consequently, they did not accept the corresponding myth, or the concept, of power.

LANGUAGE AS POWER

Power is conveyed through a variety of modes and vehicles. One of them is language. Within the Western culture, and perhaps throughout all cultures, language has been an influence and power second to none. The legendary power of language, ascribed to it by primitive societies, does not belong to the past. We are the language animal. Many of our present transactions and trading in power occur exclusively through the medium of words.

The symbolic or sacramental aspect of language makes sense of our cultural and spiritual strivings. The history of human culture and our quest for spirituality (we can replace 'spirituality' with 'humanity' for the sake of those who are allergic to the remnants of past religion) would have been inconceivable without language as the symbolic conveyor of the whole range of values. In most human transactions, whether we persuade or coerce, whether we make promises or make excuses, whether we make threats or ensnare another person in a subtle web of compliments, the power of language is as potent as it is hidden; and it is a real power.

It has been a tragedy for 'developing' countries – indeed, for all countries of non-Western civilizations – that they accepted not only Western science and technology, but also the mythos that goes with this science and technology. They also accepted the very language that conveys and perpetrates Western myths. The English language is a supreme expression of Western empiricism and, more recently, of Western scepticism and materialism. What I wish to suggest is that our problem with ourselves, with nature, and with the notion of power itself has been intensified because the English language – increasingly our tool and vehicle – seems to be more loaded with the Western secular expansionist ideology than any other language. After all, it was Bacon who conceived of knowledge as power; it was Hobbes who advocated that man is a wolf to man; it was Adam Smith who eulogized the free market economy; it was Darwin who wrote the *Origin of Species*. Bentham and Mill were the peddlers of utilitarianism in the field of ethics, which later became a justification for quite a variety of philistine values. Given the propensity of the Anglo-Saxon race towards the mundane, the empirical, the soulless, and the manipulative, the dice of the English language has been increasingly loaded in favour of viewing the world as a quarry of natural resources, and in favour of viewing power as the naked force of domination.

Had Sanskrit – the language of wisdom – become the universal language, we would have had much less of a problem with power and with ourselves. Even such a provincial language as Polish would be less harmful than English, for, in spite of the fact that it was profoundly influenced by Latin, the Polish language has retained the Slav soul. The soulless English is the source of our trouble.

The Western world has been defining, formulating and determining the rest of the world through the power of its language. The Western world has defined as primitive that which precedes the Western scientific civilization. There is quite a variety of pejorative connotations concealed in the word 'primitive'. To be primitive is to be backward, almost half-human; to join the West in its quest for progress is an imperative, an advancement, and almost a necessary condition of being human. Spellbound by Western definitions, Third World nations dance to the tune of industrial music and indeed consider themselves primitive. They see themselves as Western imperial language wants them to see themselves, not as they actually are within the fold of their own culture.

It is the power of language and the power of definitions that so often makes the Third World nations more helpless than they are; and it is this that makes the industrial nations much more powerful than they actually are. The very language that we use spells out the mastery of the West, and the submission of non-Western countries. *For there is white supremacy built into Western language. In the power of this language there are concealed many injustices and inequities.* And this power of language is almost magical and can hardly be resisted by Third World nations, in spite of their better judgement and in spite of their culture. Their heritage and tradition have become emasculated because Western modes of thinking, as expressed by English language, have become the basis for their reasoning and for their criteria of judgement.

The conclusions following from my examination spell out a good augury for Third World nations – but not in the short run. This process of rethinking of the present paradigm of power should inform Third World nations that within their heritage there are tools, concepts and traditions which may enable them to create a new context. Out of their old cultural heritage and old mythologies, which are by and large symbiotic and holistic, a new context can be created whereby new civilizational perspectives may become a reality. In a fundamental sense, the creation of new paradigms – whether of society, or of power, or of alternative lifestyles – will be easier in the Third World because the influence of the Faustian ideology is often only skin deep there, while it has pervaded all the recesses of our mentality in the West.

Working out new forms of life is never easy, anywhere. As Third World nations are locked in the cycles of the struggle for power exactly on Western terms, they only perpetuate the Western disease. As the Third World nations are attempting to fight against the West with the weapons of the West, it bodes ill for them, for they have chosen the weapons of the master to fight the master, while they are only novices.

The power to define is the power to control. The power to control is the power to define. By defining we are not only describing; we are also judging, asserting values, controlling, exerting subtle and continuous influence on people's minds. When we are pinned down by someone else's language to the point of destruction, we must get out of the clutches of this language and of the entire context.

But there is also the reality of love. And there is the language of love. When they are rediscovered, they will inform us of another reality of power. In human affairs we cannot attain justice and equity, and indeed peace among nations, without bringing back love as a reality of our existence. Our language is at present so washed-out that we do not even dare to talk about love – a sad comment on our perverse conception of the world. Love is one of the great treasures of the human species. We must not abandon it so that it does not abandon us. When people become wiser, love will assume its rightful place as an adviser on all things, including power. No doubt some tough-minded rationalists will consider this last paragraph on love as an intrusion. But why? Is not love part of the empirical world as we experience it?

TOWARDS A NEW PARADIGM OF POWER

The concept of power is a part of the complicated web of relationships and ideals that have contributed to the making of the Western mind. The myth of power is embedded in the myth of the good life. When the transcendental eschatology was abandoned and secularism began to triumph as a total philosophy, this gave rise to a new concept of salvation: fulfilment here on earth, which was later translated into the ideology of instant gratification.

Secularism, combined with material progress, was supposed to deliver the desired fruit and would make the good life a reality. The development of high-power technology as well as objective, clinical, atomistic science have been hailed because it was assumed that the more powerful the tools of physical transformation, the better off we would be in the scheme of material progress.

Thus the ideal of power, as the capacity for physical transformation, was enshrined and embedded (if only subconsciously) in the whole context of the ideology of secularism. In short, secularism, when sufficiently spelled out, resolves itself in the myth of physical power. Let us review the various components making up the myth of power.

Although the myth of power in our diagram is located on the lowest level, it really pervades all levels; it is influenced by each level and influences each level. There is, in fact, a feedback relationship between each and every level; they support each other and define each other. The diagram makes us also aware that the worship of power is not accidental but essential in Western society. The civilization that conceived of the world as a mere physical aggregate of objects to be exploited for the sake of its sensuous gratification was bound to arrive at the worship of brute power. The celebration of the ruthless is laid in the very blueprint of secularism.

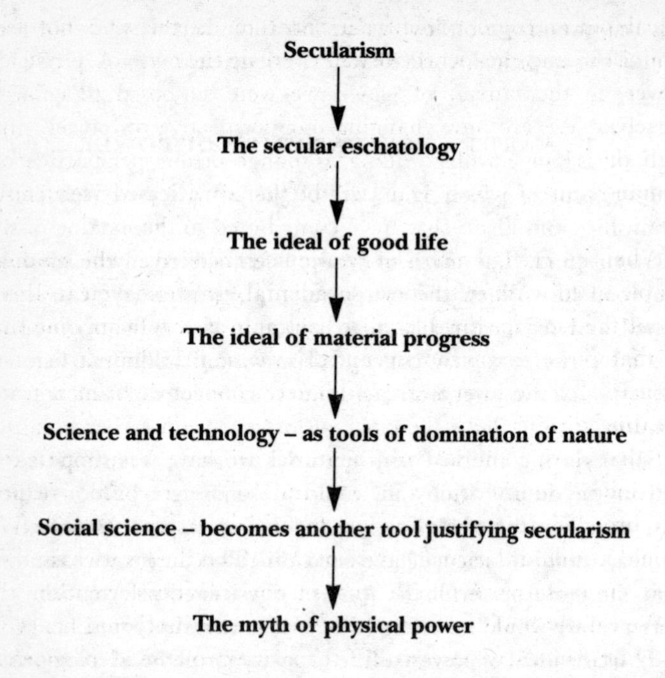

If so, why did the pursuit of naked power not become total in Western society? Because secularism, as the ruling ideology, has not been universally accepted. The ideals of our earlier creeds have consistently modified our Faustian quests. These older ideals were often based on transcendental conceptions of the human, which sought fulfilment in terms other than a mere material gratification.

Our diagram also shows that we cannot hope to 'tame' power in its present embodiment by addressing ourselves to power alone, and in isolation from the framework that has generated it. It is this larger framework which was once considered a great achievement of Western civilization that is now turning out to be such a liability. It is again our cosmology that is backfiring.

We have created the context that has elevated the machine and correspondingly dwarfed the human being. We have created the context that has impoverished human nature so that mechanical power could prevail. In order to do this and not feel diminished and cheapened we had to create the myth of (physical) power, in the pursuit of which we were supposed to enlarge ourselves. We are now changing our perspective on power, and with this change will go along a change of our perspective on human nature – which again will be vested with powers intrinsic to it.

What are, then, the new civilizational perspectives which would enable us to work out a new paradigm of power? We can learn about these perspectives looking back into history which informs us that power as authority and power as extraordinary inner capacities of the human are intimately connected. I am not advocating turning back the clock of history. I am rather pointing out that during the last two centuries we have been overtaken and indeed squashed, by the cultural counter-revolution which has brought about vast social and spiritual devastations. We should mend our language and not talk about the last two materialist centuries as a revolution but as a dark reactionary epoch.

We cannot hope to 'tame' power or use it wisely and benevolently as long as we are within the context of the ideology that has made of power the instrument of physical domination and cultural suppression. We must therefore work out a new sociopolitical paradigm. We have to create a new social reality, which will provide a new context for an alternative paradigm of power. It will not be sufficient to replace capitalism with communism, or with any form of socialism, for (as I have argued) industrial socialism, and indeed the entire Marxist ideology, are based on the conception of power as domination, class struggle, antagonism and warfare. Therefore, to replace the sociopolitical context of capitalism with the sociopolitical context of Marxism would mean replacing one vicious context of power with another vicious context. In any case, Marxism as a possible ideological alternative has collapsed. What we must do, therefore, is to go deeper, to the very foundations of our civilization, to the very roots of the Paracelsian–Machiavellian–Baconian–Galilean–Faustian–Marxist–Leninist–Taylorian tradition.

This may appear to be a tall order to consider 'realistically'. Whoever seeks more immediate and easier solutions is not a realist, however, as he or she does not perceive that the roots of our present concept of power go very deep and are now embedded in myths. In the process of demythologizing the present concept of power many fundamental relationships will have to be redefined. We shall need to redefine progress and the very idea of the good life; we shall need to re-establish organic diversity as the *modus operandi* of life and work. It is through active engagement on various levels of life that we fulfil ourselves and give meaning to our work. In the process, we shall need to work out alternative lifestyles; we shall need to enshrine frugality that will signify, not misery and deprivation, but elegant frugality. We shall need to look at human nature as an extraordinary repository of human powers. All those acts of redefining reality around us should be seen as new forms of participation in life, not as a disturbance of our consumptive slumber but as a joyous journey forward. I am aware that I have not outlined here a new paradigm of power. I have outlined the conditions necessary for it. Let others join me and articulate the details of the structure that is latent in my discourse.

Yet the rudiments of an alternative concept of power have been at hand all the time. Jesus believed in the concept of power based on love; and so does Mother Teresa. She does not make sense, 'rationally' speaking. Yet she prevails. Mother Teresa is admired by most of the tough-minded rationalists, because her love *is* power. If we had the courage to say simple things, we would say: we want to bring about *reverential power*, for this is the power that heals and builds. Reverential power requires and demands the reverential universe and reverential human beings.

One final comment: I shall not be distressed by the usual grumbling (which is too often an expression of impotence rather than an expression of a genuine search for alternatives): 'It is all very well to outline these philosophical alternatives, but tell us *how to* go from here to there.' I do not consider the question *how to* of utmost importance. This question may be of paramount importance within the universe of technicians guided by instrumental values, but not in the universe of complete human beings guided by intrinsic values.

More importantly, the 'how to' question *changes* its meaning when we are in different paradigms. What is impossible within one set of constraints becomes entirely possible within a new context. Our quest is to create a new context which will make us masters of our destiny, and not numb slaves of a blind overpowering leviathan. The project is not a new one. Human life is the story of liberation from the constraints that we have inadvertently imposed upon ourselves. When we find our new vision compelling and necessary, we shall find means for bringing it about. We shall create the new context. We shall find inner powers within us which will not only testify to the new (and at the same time ancient) reality of human power, but which will also be a fountain of new strength in rebuilding the social reality.

CONCLUSION

Power resides in the cosmos. We are repositories of this power. This suggests a profound unity between the universe and ourselves – as we recognize ourselves as the extension of the powers of the universe. Eco-philosophy provides the framework for the analysis of this unity. Actually, the *raison d'être* for spelling out our relationship to the cosmos, in terms of power, is given in chapter 4. We are woven into the cosmos; we see ourselves as its guardians, as its receptacles, as its beholders, for the sake of furthering the universe. The evolutionary concept of power here outlined may be considered the eighth pillar of eco-cosmology. How should we use the power that resides in us? Not for destruction but for transcendence, not for diminishing life but for its celebration. Let us be mindful once again that reflection and action are closely tied together: as we understand the nature of action so we act upon it. To use power intelligently we must first understand deeply its extraordinary complexity.

7

Space, Life and Modern Architecture

THE HUMAN AS A SPATIAL ANIMAL

Frank Lloyd Wright has once said that he could design a house in which no married couple could survive for more than six months, and he could guarantee a divorce as the result simply of the spatial arrangements of the house. This may be the overstatement of a supremely arrogant architect. However, it is also a statement expressing a deep understanding of how extraordinarily subtle and complex are the properties of the spaces in which we live.

In a different context another architect, Louis Khan, has said that, 'The right thing badly done is always better than the wrong thing well done.' These two opinions of Wright and Khan provide us with a point of departure that will enable us to reflect in depth on the dilemmas of modern architecture. We as a civilization are in possession of an extraordinary array of technical means. We can do, and often do do, things exceedingly well, which, alas, should never have been done at all. We are simply possessed by our means, whose satanic grip eclipses our perspective on the ends. A civilization dazzlingly rich in means and poor in ends is what we have become. In so far as our status as human beings is defined by the ends we aspire to and identify with, we are short-changing ourselves as a race. This is strikingly visible in our architecture.

Our skyscrapers and tower blocks are built with technically superior means but they will not survive, for they exemplify the meanness of human spirit and, so often, the poverty of human imagination. 'Doing more with less' has been the dominant motto of twentieth-century architecture. Looking at the situation critically, we can see that the opposite and paradoxical has happened. In comparison with other epochs, which have left behind splendid architectural monuments, built with very limited technical means, we seem to be bent on accomplishing the reverse: with technically dazzling means, we are destined to leave very little behind. Thus twentieth-century architecture could be characterized as 'doing less with more'. We are a peculiar culture: so rich in means and so shoddy in lasting results.

Let me emphasize that although space is the subject of ceaseless preoccupation of architects, it is ill-understood by them; or should I say, it is only superficially comprehended and cared for, mainly because architects are forced to limit their thinking and designs to geometric space and physical space. They are neither trained nor encouraged to care for other aspects of human space: social, psychological, cultural or spiritual.

The extraordinarily high rate of divorce, as well as the variety of other social and individual ills in the US and throughout the Western world have a variety of causes. One of the most important, in my opinion, is the senseless and often brutal way in which architects and builders go about designing and arranging living space. When cities have become concrete jungles inhabited by motor cars, when deadly suburbia further saps human energies and castrates human spirit, and when within the walls of homes you are continually at the mercy of mechanical gadgets and stupefying TV, which further disengages you from the process of life, then you must expect nothing but trouble. An unsatisfactory arrangement of space is absorbed and interiorized by the individual and often resolved indirectly through violence, frustration and depression.

For the human is a *spatial* animal. Evolutionary and social development has endowed the human organism with a number of spatial needs. One group of such needs is the unstructured environment of forest, of mountains, of open fields, in which the individual can roam and in which the visual vistas are unconstrained, so that the eye can meet a variety of natural and amorphous forms. This need is a consequence of the period when we lived as hunter-gatherers; and no doubt the consequence of earlier periods of our evolutionary history.

Another group of spatial needs is the social environment – an environment dense with social interactions. This is the consequence of our social and cultural history. There are other spatial needs, as related to sacred spaces, of which modern architecture is almost oblivious. Let it be emphasized that the need for unstructured spaces and for social spaces is almost universally recognized, yet the prevailing system of building seems to be oblivious to these deeper human needs. The game is about cost benefit analysis, maximization of efficiency, and other 'objective things'.

Within the technological system – and its imperatives of objectivization, quantification and standardization – we have reduced the variety of spaces to purely physical, Newtonian space. Being only dimly aware of this, we have been forced to design Newtonian space, which recognizes extent and volume but does not recognize non-physical qualities and attributes.

The concept of technological space is a variant of Newtonian space. Technological space attempts to arrange environments according to the dictates and demands of the industrial system and efficiency. Major environmental and social calamities have occurred in recent times because we have uncritically accepted technological space as the basis of our design activities. The limitations of technological space are obvious to anybody who stops to reflect upon its characteristics. But only recently have we allowed ourselves the luxury of such a reflection. As a result, we have reintroduced into the language of architecture other aspects of human space: social, psychological, spiritual and aesthetic. These are sometimes covered by one name – *existential space*.

Recognition of these aspects of human space is nothing new. The builders of the Gothic cathedrals were living embodiments of this recognition. But we need not go back so far; we can see this recognition explicitly expressed in our own day among the American Indians, particularly the Plains Indians.

Black Elk Speaks is a moving account of the last days of the Oglala tribe of the Sioux nation. Black Elk is a holy man of the Sioux. When he bemoans the destruction of his people, he refers not only to the physical extermination of the Sioux nation, but also to the destruction of the space that was so important to the well-being of his people. The circle was the sacred shape in the Sioux system of beliefs.

> The life of man [says Black Elk] is a circle from childhood to childhood and so it is in everything where power moves. Our teepees were round like the nests of birds and these were always set in a circle, the nation's hoop, a nest of many nests, where the great white spirit meant for us to hatch our children. But the Wasichus (Whitemen) have put us in these square boxes. Our power is gone and we are dying, for the power is not in us any more. You can look at our boys and see how it is with us. When we were living by the power of the circle in the way we should, boys were men at thirteen years of age. But now it takes them very much longer to mature.[1]

Whether we are willing to accept the mythology of the Sioux, or of the medieval masons for that matter, is of secondary importance. What is of primary importance is that we should realize that both medieval builders and the Sioux impregnated their conception of space with transphysical characteristics. This conception of space was an intrinsic part of their conception of themselves and of the world at large. *Our conception of space is a function of our culture.*

In order to envisage and design spaces that satisfy the variety of human needs – with aesthetic and spiritual needs perhaps the most important – we may have to recognize the human being as sacred. We can perhaps say without exaggeration that it is within sacred space that the quality of life resides. Anyone who is allergic to the term 'sacred' because of its associations with past religions can use the term 'existential space' to denote its peculiarly human qualities, its irreducible and specifically human content: aesthetic, spiritual and cultural.

Quality of life as a product of a person's interaction with their environment does not stand a chance in the sterile geometric spaces of modern architecture. Life does not like to be boxed. Life likes more amorphous, varied spaces. Our biological heritage is more attuned to nooks and crannies, the irregular and the round, than to linear geometry. In actual experience we find linear cities and other habitats expressing the canons of geometric planning unsatisfactory, if not disturbing, because they do violence to our biological heritage, to the amorphous and irregular in us, which is the stuff of all organic life.[2]

Architects and designers must not be afraid of the idea of life, and the criterion of the quality of life. Architecture is about life. Our contemporary social rebellion against the sterility of modern architecture is an expression of these judgements of life. Since we cannot and should not avoid them, we must attempt to anticipate and meet them. Quality has become a prominent term in recent architectural discussions. The reintroduction of the concept is an acknowledgement that judgements of life are once more relevant to architecture.

FORM FOLLOWS CULTURE

Architecture recapitulates culture, of which it is a part. In a flourishing culture, architecture partakes in its glory. It then expresses not only firmness and commodity but also delight. When a culture is decaying and unable to sustain its idiom, architecture comes in for much of the blame because its shortcomings are strikingly visible and experienced by all. While other social and political institutions, including educational ones, can more readily camouflage the malaise of the culture which is expressed through them, architecture conspicuously reflects both triumphs and shadows.

Theory and practice are intimately connected. Immanual Kant said that theory without practice is impotent and practice without theory is blind. Much of our present practice is blind because it is not informed by theory or rests on non-viable theory. The fusion of theory and practice is particularly striking in architecture. Architecture constitutes a bridge between logos and praxis; it is a point at which the two meet. For this reason architecture *visibly* demonstrates the greatness of our visions and equally the failure of our large conceptions. In architecture, in brief, many of the ideas discussed in previous chapters find a visible expression.

The architecture inspired by the mechanistic logos has demonstrably failed us. The deficiencies of present architecture and its inability to shelter us adequately and to provide spaces that are life-enhancing is not so much the fault of architects and builders, but the fault of those larger conceptions on which architecture and our culture are based. It is at this level of analysis that the relevance of eco-philosophy is second to none: for it helps to understand in depth the deficiencies of present architecture and it indirectly provides a new foundation for architecture and the design process.

We have to change our logos so that our practice is not blind or counter-productive – either in architecture or elsewhere. Architecture, indeed, should be seen as a symbol of the multitude of all other activities in which counter-productive practice is a result of the mistaken logos. We are back to our diagram linking cosmology with action (see p. 12).

The shadows of twentieth-century architecture are now very conspicuous – so much so that society is alarmed by them and architects are themselves deeply disturbed by the state of affairs. To heap scurrilous abuse on the products of contemporary architecture is as easy as it is futile.

Now, while it is a cliché to say that society has the kind of architecture it deserves, it may be less of a cliché to maintain that society has the kind of *culture* it deserves. And it may not be a cliché at all to suggest that there is an intimate correlation between architecture and culture, that, in general, architecture is a function of the dominant culture. At times, on the other hand, architecture significantly articulates and helps to define general culture. This dependence of architectural form on the ethos of a given culture can be seen in all cultures, including 'primitive' ones.

In the culture of the twentieth century, Western architecture is dominated by, and defined through, economics and technology. If we look with a discerning eye at the variety of so-called new trends and tendencies, we cannot help observing that almost all of them are an expression of the technological ethos; epicycles of the technological system. Brutalism or Venturism, Archigram or rational architecture, operationalism or new rationalism – all bear the technological stamp. They are products of twentieth-century technological society. The Bauhaus (particularly in its classical period, when it became the dominant architectural ideology) epitomized the technological apologia. It found expression in the cult of efficiency and functionality, the belief in the machine and standardized norms, the worship of new materials and techniques. To these, all other aspects of architecture were to be subordinated.

It is not that we want to build sterile buildings, shoddy environments, spaces in which the human spirit is thwarted; our culture *makes* us design such environments and such spaces. There is something insidious in the spectacle of dedicated, determined and talented architects who can, and want to, build much better than they are allowed to or can afford to in the present context.

Building regulations and prohibitions have always existed. In the past they acted as filters through which the best of human imagination penetrated to produce the finest achievements. Now those filters have become so monumental that they do not let any imagination pass through; they allow no quality to emerge at the other end. This, of course, must produce a feeling of frustration, if not a slight schizophrenia, in the architects who utilize so much energy and imagination at their end and see so few positive results at the other. The best of their imagination is stuck in the filters and eliminated in the process.

Now, the power of those filters is not accidental; it is essential to the way in which the quantitative technological society works. It is not that we have incautiously multiplied codes and regulations, which are getting out of hand. It is rather that those codes and regulations guard and enforce the ethos of technological culture.

We have created a culture that systematically destroys quality. Those 'monumental filters', which interfere with our imagination and our desire to produce quality environments, are the culture's protective devices working to bar products which go against its quantitative ethos. Why do we then build inadequately? Because we have an inadequate culture. The culture is the filter.

Our growing ambivalence towards and sometimes outward rejection of planning is significant and symptomatic, not only because it demonstrates some important shifts in architectural theory and in articulation but also because it signifies important shifts in our culture. The rejection of planning is an implicit rejection of the linear, mechanistic, geometric, predominantly logical and economic mode of thinking and acting in favour of more organic, intuitive, decentralized, ecologically sound and life-enhancing forms. In changing our hearts and minds about planning, as a viable and necessary vehicle of architectural theory and practice, we simultaneously (although indirectly) change and re-articulate the idiom of our culture: away from its mechanized, objectivized matrix and towards one that accommodates and accentuates quality.

The architectural dogma of the first part of the twentieth century was: *form follows function.* When function became limited to its physical and economic parameters, we, as human beings, found the resultant form constraining and suffocating. So the slogan has been quietly dropped. 'Form follows function' was a specific articulation of technological culture within the realm of architecture. With hindsight we can now suggest a much more adequate characterization of architecture: *form follows culture.* Alternatively we could say: *shell recapitulates spirit* (i.e. as expressed by a given culture), or even *shell accommodates spirit.* Thus, neither form follows function, nor shell before performance, but appropriate form to accommodate the spirit of a culture. In short, *form follows culture.*

Cultures always seek to vindicate their specific tenets and achievements, even their obviously pathological traits. The shortcomings of our own culture are painfully reflected in the architecture of the recent past, which simultaneously signifies a technological triumph and a human plight: that architecture constitutes clear evidence of the deficiencies of our culture with its glorification of the objective, the physical and the efficient and its attempts to diminish the spiritual, the sensitive and the humane. In a nutshell, significant changes in architecture are not going to be accomplished either by introducing more efficient technologies or by simply manipulating architectural theory. If we wish to change architecture we cannot limit ourselves to architecture or start with it alone. We have to start with – or simultaneously address ourselves to – another level, the level of the general culture that underlies the thinking and behaviour of the age we live in. That start has in fact been made, and we are already changing the idiom of our culture and the idioms of architecture and planning – away from the linear and pseudo-rational and towards the ecological, the organic, the compassionate. In so far as we are doing that, in a muted and tentative way still, we give testimony that we are prepared to change our world-view, and thus our cosmology.

THE QUEST FOR QUALITY

Quality is a difficult term to define, particularly in the context of the quantitative society. Moreover, the idea of quality in architecture is by no means a simple one. The term can signify at least four different though interrelated things:

Q_{I} Quality of design (the original idea and its representation on paper, i.e. the drawings).

Q_{II} Quality of processes (the various means, techniques and technologies employed in the process of construction).

Q_{III} Quality of products (the assessment of buildings upon their completion).

Q_{IV} Quality of life generated by built environments (the quality of people's interaction with environments).

It is, of course, the fourth level that is all-important because it ultimately redeems all the others. Ideally one would like to assume that quality on one level implied quality on the next, so that quality of design resulted in quality building, and ultimately in quality of life in the built environment. In the real world the relationships are not that simple. They are clearly asymmetrical. Quality at the first three levels: design, processes and building, does not automatically guarantee quality at the fourth level – the 'quality of life' of built environments. Thus the conjunction of Q_I, Q_{II} and Q_{III} does not automatically imply Q_{IV}. Let us put it in the form of a formula ($\nrightarrow$ means 'does not imply that'):

(a) $(Q_I \cdot Q_{II} \cdot Q_{III} \cdot) \nrightarrow Q_{IV}$

There are other asymmetries. Q_I does not automatically secure either Q_{II} or Q_{III}.

(b) $Q_I \nrightarrow Q_{II}$; $Q_I \nrightarrow Q_{III}$, and furthermore:
(c) $(Q_I \cdot Q_{II}) \nrightarrow Q_{III}$

We can assess quality only *post hoc*, only a posteriori. If we find a given environment working, if it generates interactions that we deem to be life-enhancing, then we judge the building to be a success. Quality of life, Q_{IV}, automatically implies Q_{III}.

(d) $Q_{III} \leftarrow Q_{IV}$

A good example is the National Theatre on London's South Bank. Although many have questioned the processes and the product, those cold, brute masses of concrete, the quality of life that the environment of the National Theatre has generated is such that we have to consider the building a success. So, quality of life generated by built environments automatically secures all other qualities.

(e)
$$\left.\begin{array}{l} Q_I \ \nwarrow \\ Q_{II} \ \leftarrow \\ Q_{III} \ \swarrow \end{array}\right\} Q_{IV}$$

The quality of the final link, of the culminating point, determines the quality of the contributing links. There is another justification for this criterion of quality. In logic, there is a law called the retransmittability of falsity. If p implies q ($p \rightarrow q$), then it does not automatically follow that q implies p ($\bar{q} \rightarrow \bar{p}$). But it does follow that if not q, then not p ($\bar{q} \rightarrow \bar{p}$); in short: $(p \rightarrow q) \rightarrow (\bar{q} \rightarrow \bar{p})$. In other words, a negative judgement about the quality of life generated by a given environment (building) automatically retransmits that negative judgement to the contributing links Q_I, Q_{II}, Q_{III}, either to some of them or to all of them.

The dominant mode of thinking has been to concentrate on Q_{II}, quality of processes, for this is the particular quality which is favoured by the technological system. Quality of processes can be sub-divided into at least three sub-categories:

- Adequacy of technical means
- Adequacy in meeting urban and planning requirements
- Adequacy in meeting economic criteria

The search for quality in our day is clear proof of the fact that we have not lost the sense of the concept and, moreover, that we are transcending the boundaries of technological culture. This search for quality can be seen in unlikely places. Robert Pirsig's book, *Zen and the Art of Motorcycle Maintenance*, is relevant to our discussion here:

> Any philosophical explanation of Quality is going to be both false and true precisely because it is a philosophical explanation. The process of philosophic explanation is an analytical process, a process of breaking something down into subjects and predicates. What I mean (and everybody else means) by the word quality cannot be broken down into subjects and predicates. This is not because Quality is so mysterious but because Quality is so simple, immediate and direct . . . Quality cannot be defined . . . If we define it we are defining something less than Quality itself.[3]

Quality-of-life architecture is architecture which has the courage to recognize the spiritual and transcendental dimensions of the human being. Quality resides in spaces that are deliberately and purposely endowed with characteristics and attributes that are transphysical. Physical well-being is not a physical state but a psychological one.

The two outstanding architects of the twentieth century, men who held life at the centre of their vision and designed to meet the requirements of the radiance of human life, have been, in the first half of the century, Frank Lloyd Wright and, in the second half, Paolo Soleri. While Wright designed exquisite *individual* houses which blended organically with their environment, Soleri has designed whole cities, called arcologies (a fusion of architecture and ecology), which meet the challenge of the post-industrial era by attempting to blend man and nature in a novel way. Soleri's arcologies are monumental architectural designs for frugal, ecological living. At their centre, however, lies the human imperative, which demands that, in addition to economic and physical needs, the desiderata of human life at its most developed cultural and spiritual levels shall also find full satisfaction. Soleri's exemplary achievement is to have expressed, in his arcologies, the architectural, the ecological, the frugal and the spiritual.

In summary, the true purpose of architecture is to continue, enhance and celebrate life. The phrase 'to continue, enhance and celebrate life' must be seen in its proper context. The industrial sharks who destroy our ecological habitats for profit and often force architects to design anti-life environments can claim to be continuing, enhancing and celebrating their own lives. Individual greed must not obscure from our view the fact that the eco-system urges various constraints upon us. In addition to our ecological awareness we must have a coherent conception of the human being and a viable model of culture that are capable of sustaining us in the long run.

In our lowbrow culture, which is so often proletarian in the worst sense, the architect must assert his role as a patrician, must lead instead of bowing to acquisitive and materialist preferences.

Only when people transcend their obsession with material acquisitiveness – which is one of the chief causes of environmental destruction and of our inner emptiness – will it be time for the architect to relinquish the role of the designer of complete environments. This may be an unpopular suggestion, contravening as it does the ethos of egalitarianism. But have we not had enough mediocrity, environmental débâcles and disasters to realize that the egalitarian ethos is not capable of designing, in this complex world, to meet quality-of-life criteria? For the egalitarian ethos (or the anti-élitist stance) too often tends to be standard, undistinguished, careless and morbid, thus ultimately leading to anti-quality spaces. Thought as revolution means questioning every cliché of our times, which may be well intended but which nevertheless may result in banality, mediocrity and bowing to the lowest common denominator.

The reorientation of our vision concerning priorities calls for a new idiom of architectural thinking and articulation. It also calls for a redefinition of the present idiom of culture. The new idioms of architectural articulation and of culture will parallel each other because they will be two different aspects of the same thing.

We are clearly groping towards a new paradigm in architecture. I use the term paradigm in the sense in which Thomas Kuhn uses in *The Structure of Scientific Revolutions*: as an overall

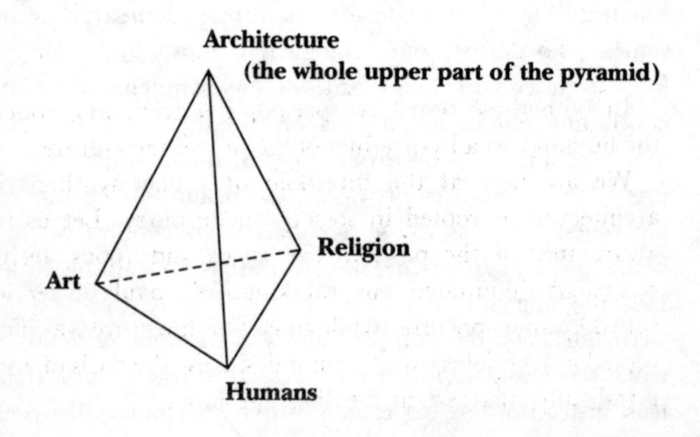

conceptual umbrella which includes assumptions, practices, theories, as well as criteria of judgement of finished products. By shifting the emphasis from technical virtuosity of means to the quality of life, we are changing not just a small part of practice; we are changing the whole idiom of architecture, we are changing the entire paradigm.

Traditionally through millennia, architecture was defined by a context in which religion and art predominated.

In the nineteenth century, and particularly in the twentieth century, we changed the context and architecture came to be defined and determined by economics and technology.

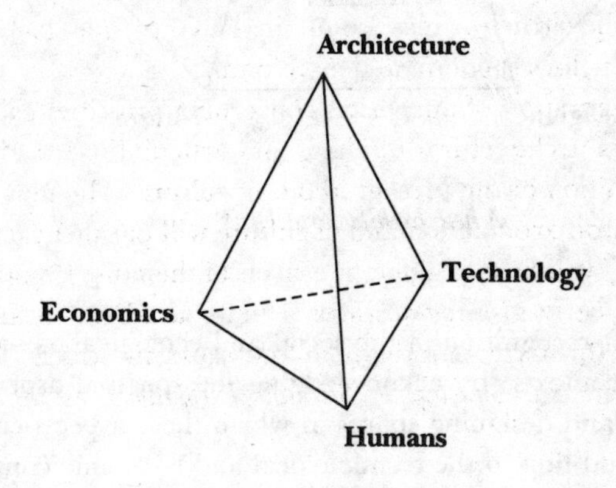

In both these respective periods the prevailing conception of the human was a by-product of the prevailing culture.

We are now at the threshold of a new synthesis in which architecture is rooted in society and ecology. Let us be clearly aware that in the past, in the 1950s and 1960s, technological escapades dominated our thinking. We paid lip-service to so-called human concerns, while in fact architecture was increasingly rooted in technology and economics. Now the basis of architecture is radically enlarged, as the diagram shows.

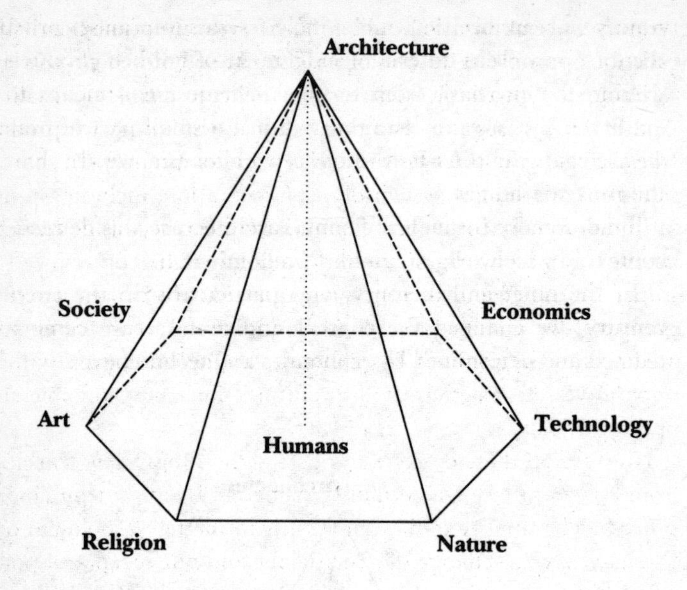

A new architectural basis

Our explicit recognition of the social and ecological as well as the religious contexts (by acknowledging the spiritual aspects of our existence and designing spaces in which those aspects can be fulfilled), in addition to the technological and economic contexts, makes us look at architecture and design in an altogether new way. The new paradigm is in the making. This new paradigm is evolving around the new imperative that the true purpose of architecture is to continue, enhance and celebrate life. This imperative, as we can clearly see, is an extension of the ecological imperative we discussed in chapter 4. This is to be expected: if the ecological imperative is valid it should be applicable to a variety of fields. Architecture is one of them.

The quality-of-life criterion, which is the architectural formulation of the ecological imperative, has a number of specific and tangible consequences. Quality means breaking away from the tyranny of centralization, which is the tyranny of mediocrity. It calls for appropriate technologies, usually soft (for high-powered technologies invariably destroy the delicate tissue of quality). And it implies frugality and durability in the things we produce (the throw-away society is the arch-enemy of quality). In short, it signifies a whole new system of instrumentation, including a new attitude towards work – you cannot have a quality-of-life environment if human work is systematically degraded or reduced to stupefying, mechanical, repetitive tasks. These consequences of quality-of-life architecture are too fundamental to be accommodated within the confines of today's technological culture with its imperatives of objectivization, quantification and economic sub-optimization.

How are all these changes to be brought about? The transition from any form of logos to daily practice is always a troublesome process. The quality-of-life criterion is ultimately a political one. We may have to change our politicians and our social and political institutions before this criterion can prevail. We shall need people with superior judgement, at various levels of the political hierarchy, who can see that quality-of-life environments are not a luxury but a necessity: environmental, social, ecological, psychological, aesthetic and political.

That architecture is a global phenomenon, cutting across various layers of life and involving everyone, there can be no question. The involvement of Prince Charles in the debate over the nature and the state of architecture, during the late 1980s, is eloquent evidence that architecture belongs to all. Prince Charles is not an architect, but he is a sensitive and responsive human being. He protested against the soullessness of modern architectural monstrosities in the name of the quality of life. His points were not original. But they were nevertheless valid and well received. Many ordinary people felt that *if the Prince said so, it was all right not to be intimidated by the machine* and, moreover, it was all right to express one's intuitive feeling about what good architecture should feel like. *In the final reckoning, the argument was not about architecture but about the shape and meaning of life.*

Architecture has a great and salutary role to play in the future. Society expects architects to take care of the complete environment, for there is nobody else to do it. Architects cannot refuse this responsibility. In recent years they have gone halfway towards their destiny by designing new blueprints for alternative habitats, including new uses of solar energy and frugal management of scarce natural resources. In order to fulfil the social responsibility that is thrust upon them, architects must have the courage of their convictions, to prevail with their quality criteria over mountains of paralysing regulations, and to be relentless and unbending in bringing architecture back to life.

We must not blame everything on 'the system'. This would be too easy. We must see ourselves as a part of the equation. We must infuse ourselves with the spirit of innovation, inspiration, oneness with the environment and compassionate caring for others so that it shines through our designs and buildings, and the system is transformed by our acts of creative transcendence.

BEYOND THE MAGIC OF THE MACHINE

It has been said of our times that 'the machine is a metaphysical symbol of the whole age'. This is more true of the recent past than of the present. The machine no longer casts a magic spell on our minds. But it was precisely because of this spell that some eminent architects were proud to define the house as 'the machine for living'. Buckminster Fuller was one of the first to so define the house. Because of this infatuation with the machine, with technology, with the objective hardware, we were happy to reduce knowledge to information; and the enormous variety of acts of understanding contained in the design process were reduced to objectivized design methodologies. We are now in revolt against the culture of the machine. And many of the pronouncements of the architectural prophets of the former epoch strike us as ludicrous, if not dangerously insane.

Thus Le Corbusier wrote with conviction and a poetic flair in 1927:

> We must have the courage to view the rectilinear cities of America with admiration. If the aesthete has not so far done so, the moralist, on the contrary, may well find more food for reflection than at first appears.
>
> The winding road is the Pack-Donkey's Way, the straight road is man's way.
>
> The winding road is the result of happy-go-lucky headlessness, of looseness, lack of concentration and animality.
>
> The straight road is a reaction, an action, a positive deed, the result of self-mastery. It is sane and noble. A heedless people, or society, or town, in which effort is relaxed and is not concentrated, quickly becomes dissipated, overcome and absorbed by a nation or a society that goes to work in a positive way and controls itself.
>
> It is in this way that cities sink to nothing and that ruling classes are overthrown.

This adulation of the straight line and of the rectilinear has led to the so-called rational architecture whose final result is the tower block and the soulless box in which we, as human beings, are suffocated. In contrast to the pseudo-rationality of the straight line (the original geometry of nature is not the straight line but the curve[2]), the painter–architect Hundertwasser has declared that the straight line is obscene and it leads to hell. 'This straight line leads to the loss of humanity.' Instead he celebrates the spiral. 'The spiral is the symbol of life and death. The spiral lies at the very point where inanimate matter is transformed into life. It is my conviction that it has a religious basis . . .'[4]

We live in many geometries, and which of these is most conducive to our total well-being is an open question. The geometry of the womb is certainly different from the geometry of the computer circuit. The geometry of galaxies is more like that of the womb than that of mid-Manhattan streets. The geometry of the universe, and especially the geometry of nature – of all living forms nursed by nature – is anything but based on the straight line and the rectilinear.

Our attitude to the straight line is a dice loaded with metaphysical preferences. Casting this dice in architectural forms was, for Le Corbusier, an expression of the rational and the purposeful. Yet we now see it as creating the sterility of human spaces that appear to us as iron-clad cages. Changing our preference for geometry, while we design human spaces, is changing the nature of life lived in these spaces. The scientific mind has a preference for the straight line and for the simple geometric grid. And this mind is often imposed on our intuitive mind, which has a distinctive preference for the curve, for the irregular niche, for the primordial spiral, as Hundertwasser would put it. In our scientific age, it takes an exceptionally strong mind to resist the imposed grid of the scientific mind and pursue the intuitive notions of what the desirable human space should be. The greatest architects of the twentieth century are examplars of the intuitive mind in action. The architecture of Frank Lloyd Wright, of Louis Khan, of Paolo Soleri, is the triumph of the irregular; it is also the triumph of the intuitive; implicitly it is the denial of the validity of the geometric grid as the expression of our rationality.

Thus beware of geometry. It is not a boring subject which we have to suffer as students in high school; rather it is a subject whose right understanding is the key to the right design, a right conception of human spaces, a right idea of the quality of life.

Yet it is a fact that the bulk of modern architecture is still in the image of the machine, paying homage to the straight line. The architects of the last fifty years have been too much taken by the ideology of progress, and really brain-washed by it, so that they did not realize (until recently) that theirs was a mindless quest in search of the mechanical grail. But the grail is not a mechanical entity. The slogans of the culture are responsible for its ethos. The slogan of the 1929 Chicago Exhibition set the tone for a whole generation: 'Science Invents, Technology Implements,

Man Conforms.' For us it is now inconceivable that this kind of simple-mindedness could prevail. But it did, until we awoke from the mechanistic stupor at the end of the 1960s. This process of re-awakening is far from complete in architecture. We still produce monsters for which future generations will curse us. However, the imperative of life is gently directing us away from the sterile geometry of the linear and the rectilinear. For life is holistic. In contrast, design methodologies are atomistic. Wisdom is integrative, while information is analytical. Quality is attested by actual human experience; methodological perfection is measured by quantitative criteria. Life is diversity and life is transcendence. Architecture and built-in environments, which are going to survive and sustain human beings, must meet this diversity and, indeed, provide for this diversity.

'People need poetry,' said Ralph Erskine during my visit to the Byker redevelopment in Newcastle, England, 'and I am trying to produce it by building this habitat,' he continued. If one wants to use examples of the design process that is guided by the right imperatives, then the development and construction of the new Byker is as good an example as any, for Erskine started with the premise that what he wanted to preserve was a certain quality of life, a certain quality of social relationships.

He went to study the texture of life as was lived in the old Byker, submerged himself in the ethos of the people who lived there, and then re-emerged with a design that attempted to recapitulate life. Thus the design, the whole process and the whole product of it, attempts to shelter and nestle certain specific qualities of human existence. One of the premises was that 'people need poetry'. Let us also remember that the Byker development is not a fancy artistic community, but low-cost housing, simple council flats; and that this habitat is to shelter not just a few families, but about 9,000 people. The result is a veritable symphony of forms and colours of immense variety which, as far as one can see, are working – for life *is* diversity.

The new Byker of Newcastle is going to be enhancing and successful (I venture to prophesy), because it attempts to meet the diversity of life; it provides for diversity, is itself an embodiment of diversity. A new methodology of diversity must be created and made the basis of true planning, a grid for future human-made environments. The existing design methodologies have proved to be calamitous because they simplify the texture of life and the scope of our needs to absurdly contained conditions.

First of all we have to have the courage to recognize that our sensitivities and sensibilities are necessary and invaluable instruments of assessment and design. They are not just an expression of our subjectivity: they have been developed through millions of years of our biological evolution and through the millenia of our cultural evolution. There is wisdom in our perception and in our sensitivity, both of which have been shaped by the exacting demands of the multifarious forms of the past life of the species. We assess our environments continually through our eyes, our minds, our hearts, our guts, working in unison. Often the final assessment is acknowledged cryptically by the overall reaction of the organism which sometimes says 'yes', sometimes says 'no', sometimes says 'maybe'. Because our intuition often works in a cryptic manner many find its verdicts and assessments to be too unreliable and too subjective, and they prefer hard facts and figures. But no facts and figures can really convince you if a given environment does not look right and does not feel right. That which looks right and feels right, *is* right, for it is the response of an organism conditioned and refined by millions of years of trial and error.

As organisms in a continuous biological–social–aesthetic evolution, we are not subjective: we only express individually that which life has stored in us as the repository of its sensitivity. *Let us therefore celebrate our subjectivity and our intuition, for it is an enormous repository of the wisdom of life.* When properly handled, our subjectivity is a far superior instrument for assessing built-in environments than all those physical and quantitative criteria. Besides, quantitative criteria are established by human agents with regard to certain phenomena as experienced by other human agents. *In every act of quantitative assessment there is an a priori act of intuitive, subjective judgement*, which is built into our a posteriori 'objective' assessments. There is therefore no escape from subjectivity. But let us use it wisely: not as something that expresses our quirkiness and our idiosyncracies but as something that expresses our belonging to the sensitivity of life and of evolution. If you have no courage of your perception and your sensitivity, how can you have the courage to live?

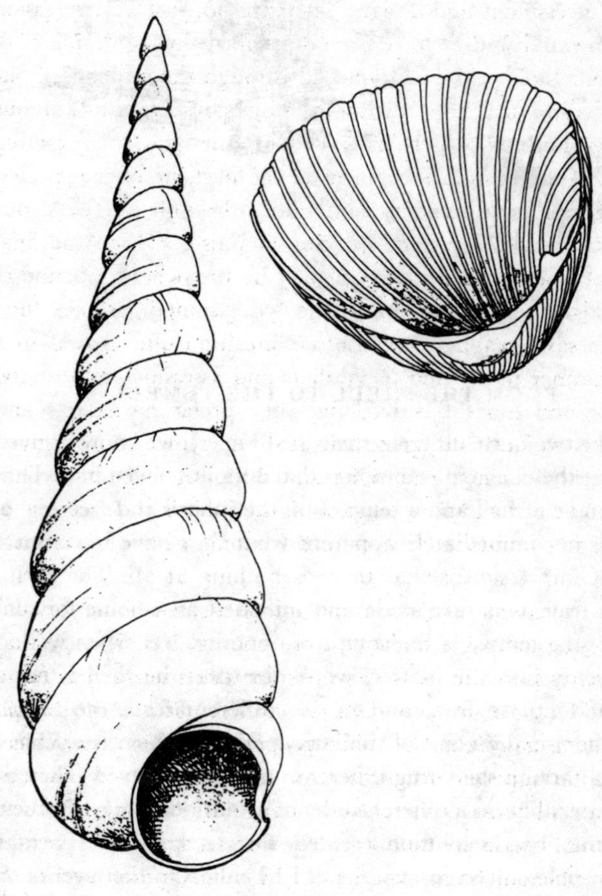

Fig. 1

Fig. 2

FROM THE SHELL TO THE TEMPLE

If we take two quite different shells (see Fig. 1), we know instantly that what they have in common is that a mollusc lived in each.

If we take a shell and a temple, on the other hand (see Fig. 2), then it is not immediately apparent what they have in common. The first impression is that there is nothing at all. The shell is small, is frail, is nature-made and intended as a home for some creature; the temple is the complete opposite. Yet when we look more deeply into the matter, we realize that the shell is not so small, and not so frail, and not so different from the temple. What, then, is the point of similarity? Each of them is a form of shelter, although sheltering different forms of existence. They are both shelters but on a different level of evolutionary development. In its unfolding, evolution created various kinds of spaces in order to shelter different aspects of its being. On the level of the human being, it created new kinds of spaces to accommodate the

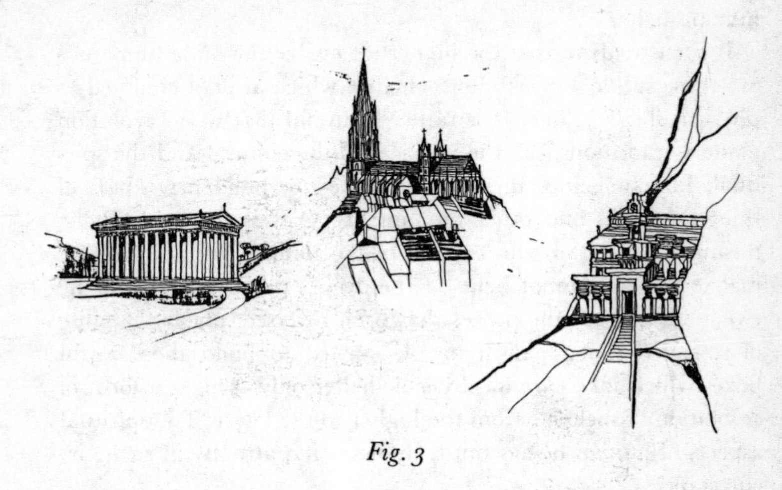

Fig. 3

increasing variety of human needs, including cultural and spiritual needs. Although they look at first so different, both the shell and the temple respond to the need for shelter, albeit on different levels of evolutionary unfolding.

Temples, whether Greek, Christian or Hindu (see Fig. 3), attempt to provide a shelter for our spiritual quests and qualities. They also make us feel at home in this large universe. They make us feel at home by relating us to heaven and earth and all. This is what good temples do, even if people are not fully aware of that. They are all shelters. But with a difference!

A shell is the original shelter. A shell represents the original geometry of life. In the exquisite beauty of the shell – in its strength and symmetry – we witness the anticipation of the structure of our dwellings and of the symmetries of our temples. The more we learn of the magic of the shell, the more capable we shall be in designing our dwellings and temples. For it is all a part of the same evolutionary rhythm – the continuous waves of the sea created the shell, then our ribs, then the columns of the Parthenon. In producing the shell, nature was already toying with the idea of the temple, and in between it created what we call the human shelter.

It is clear, then, that the higher we evolve the more numerous are those subtle aspects of our shelter, which are not confined to physical shelter alone. It is rather beautiful that when evolution made a transition from the physical to the cultural and the spiritual, human beings in response started to build new kinds of things, temples and other monuments, expressing the symbolic meaning of human culture, and of the human spirit. When we realize this we cannot help but be deeply dismayed that in the twentieth century the process has been reversed: under the guise of so-called functionalism people started to build those awful boxes which take care of physical shelter only. This is a form of *de*volution of shelter – from the higher to the lower. The spiritual aspects of human beings must, by necessity, atrophy in such circumstances.

Confined to spaces that cater for physical needs alone, no wonder people have so many problems with themselves. Evolution has required them to satisfy all kinds of forms of human existence, and they try to force it all into those soulless shelters. By eliminating mystery from architecture, we have eliminated a great deal of human content essential to human nature.

The poverty of our context may impoverish our imagination to the point where we cannot see and we then settle for the obvious and trivial, reassuring ourselves that a shelter is a shelter, that a thought is a thought – while some thoughts can begin a revolution. Everything is shelter, but on the quality of our imagination will depend whether we create shelter for the total human being or whether, by being a slave of the humdrum impoverished context, we will try to force the variety of human existence into containers that are sufficient only for the physical needs of human beings. If we follow the latter course, and my goodness, so many architects do, we will follow evolution backwards, and will become a denial of the proposition that logos and imagination can transform the world.

It is strange but simple: evolution is about understanding the variety of contexts and about understanding the increasing depths of contexts; architecture is about providing shelter that shelters all the attainments of evolution, including the higher values. *Architecture is homage paid to our understanding of the evolutionary epic, or it is nothing at all.*

When evolution created the shell this was a beautiful expression of integrity and wholeness. And so it was when evolution created human ribs, and when humans created the columns of the Parthenon. All architecture is about wholeness, integrity, complexity, becoming; also about aesthetics. In traditional art as well as in traditional architecture, the symbolism, or that which architecture and art attempted to signify, were in the service of wholeness: the symbols attempted to help us to integrate on all levels of being, particularly on the spiritual level. Technological architecture is based on the premise of fragmentation, disconnectedness, isolation, simplicity. The dimension of spirituality is lacking there; and so is lacking the dimension of wholeness, this deep structure that guides the architect's imagination to create spaces for all aspects of our existence. Because of its in-built limitations, and its accepted premises, modern architecture cannot even begin to cope with the relationships signified by wholeness and true complexity.

We cannot have architecture embodying wholeness until we have architects who are whole. At present their training goes in the opposite direction. Where and how do we establish schools for wholeness in which architects can be trained and developed in a new way? That, of course, is not an isolated question. We need universities and professional schools in which not only the wholeness of our perception can be trained but which help us to integrate as whole persons (see note 7, chapter 5).

Architecture, environment and the self are one. The motto for future architects should be: strive for your own wholeness so that your building is whole; so that you contribute to the wholeness of the environment.

CONCLUSION

The spaces that surround us indelibly mould our psyche. The struggle for a meaningful environment is one for the spaces that are empowering and not disabling; that promote and enhance life and do not reduce it to the mechanistic beat of the machine. Space in which we live is ultimately a metaphysical proposition for it is a nurturer of ten thousand things. This is the central message of eco-philosophy for architecture. We must abandon the crippling ideology of *form follows function*. Instead we need to realize that the architecture of greatness has invariably followed another principle: *form follows culture*. Our task is to create a rich context for our new culture so that architecture is not caught in the barrenness of a cultural desert. The quality of life is nurtured by existential spaces, as well as social and sacred spaces. We need to create them, or at least recreate them. We should take our clue from evolution whose context is infinitely rich and always evolving. We must strive to create architecture, and human-made environments, in the image of evolution.

8

Eco-ethics and the Sanctity of Life

SETTING THE STAGE

The human quest has been one and the same for millennia: how to secure meaning, well-being and a sense of harmony. On the lower level of human existence it is how to secure food, shelter and enough amenities for satisfying social existence. This quest, however, has been expressed differently in different epochs of history. At one time, when the environment was harsh and human tools limited, when human knowledge was in its infancy, and ignorance was in abundance, at such a time the forces of nature were seen as threatening, menacing and often malevolent. To protect ourselves from these potentially malevolent forces, we created deities and other religious symbols.

As the human story unfolds, we symbolize more and more realms of our existence. Religions, as distinct bodies of beliefs, have become articulated. The original spirituality, which was diffused throughout all of nature, is now focused. A process of transcendence has occurred whereby divinity is vested in a personal God – residing over and above the human realm. This is the story of the Judeo-Christian tradition of the West. God has become transcendent, and no longer present in all the forces of nature around us.

With regard to God so conceived, the first moral commandments were formulated – the laws binding the human to the sacred, and also regulating relationships between human beings. In the Judeo-Christian tradition, the first moral decalogue, the Ten Commandments, are mainly harsh prohibitions – 'thou shalt not'. They bind us to the transcendent God who is responsible for our vicissitudes and our ultimate destiny. Jehovah's universe is a harsh one, where people live in constant fear of the wrath of the almighty God.

In the universe so conceived, values are predominantly personal. They regulate the relationships of God to human, and of human to human. The importance of nature as a value in itself does not enter the picture because (i) nature is regarded as a threat to be contained; and (ii) because in the monotheistic Judeo-Christian tradition (which is the basis of my analysis here) all power and value is vested in a personal but invisible God, so that the visible cosmos is left disempowered.

Our cosmologies give us not only a picture of the physical universe, but also a set of values which is usually latent in a given cosmology. The Judeo-Christian cosmology-cum-theology, by giving us the dominion over everything that creeps over the surface of the earth, has articulated for us the Hebrew conception of God, and of the human as a special creature of God. It has also spelled out highly anthropocentric values, and we are still steeped in this anthropocentrism.

Lynn White has suggested that the Bible is a blueprint for environmental destruction.[1] It would be more accurate to maintain that the Bible spells out the limitations of the early monotheistic cosmology. Given the circumstances, perhaps the early prophets could not do any better. Now, with hindsight, we do know better as we are fully aware that these excessively anthropocentric values are part of our present predicament.

The aim of this chapter is not to provide a historical analysis of the Judeo-Christian tradition of values. Rather our aim is to propose a new set of values appropriate for our times and our problems. Before I propose a set of ecological values, let us very briefly look at the various traditions of values that have guided and inspired Western civilization. The three towering myths that summarize the various value traditions of the West are embodied by Prometheus, Jesus and Faustus.

The heritage of these myths is stored in the psyche of Western people. The Promethean heritage persuasively suggests that life relentlessly unfolds but not without pain. Progress is accompanied by sacrifice. The universe is beautiful but frail. If we forget about its frailty and charge like a bull in a china shop, hubris follows. To perceive the subtle limits, constraints and frailty of the universe requires wisdom. We have lost wisdom in the West, hence we have unwittingly invited the hubris which is our accompanying shadow in the present Western civilization. For the people who have got so intoxicated with power, it could not have been otherwise.

The heritage of Prometheus is noble and inspiring. We must reclaim it and incorporate it into our moral structure. The Promethean heritage is part of our overall moral heritage, as I have argued in chapter 4.

As for the heritage of the other myth, that of Jesus, we must be aware of its positive aspect. (The negative aspect is the fundamentalist preaching: 'Jesus died for your sins and you must suffer.') It is love triumphant, love that conquers all, love that transforms, love as a positive modality of human experience that emanates from Jesus's teaching. Let us notice that love is almost a spurious category within the universe ruled by the values of efficiency, control and objectivity. This is one of the reasons why love is so difficult in our times. We are not an emotionally deficient people, but our emotional expressions, including that of love, are so often blocked by a variety of objective, cold, controlling structures.

The Faustian heritage is most recent, yet it is most vocal and most aggressive. The heritage of scientific-technological values can be subsumed under one powerful myth: you only live once, therefore you live dangerously, at whatsoever expense, even if it means at the expense of ecological habitats, of other species and of future generations.

The reckless thinking of the proponents of progress in the West, their sense of conquest, their desire for power and domination, their lack of a sense of harmony, are all related to the overall Faustian drive which we have developed alongside the notion of progress. The devastation of natural environments, the brutal elimination of other species and cultures, the ruination of individual existential lives, the disappearance of social cohesion, are acts that *cannot be rationally justified*. Yet we perpetuate this irrationality daily. Why? Because of the omnipresence of the Faustian fixation: you only live once, therefore you live dangerously; you have a ball on earth, even if you leave behind nuclear deserts.

Now, in questioning the Faustian heritage we are by no means decrying all progress. Prometheus was a harbinger of progress, and what a wonderful harbinger he was! But his was a different kind of progress from our material progress.

Some philosophers and thinkers in the nineteenth century became alarmed by the collapse of religious and spiritual values, seeing them crushed by the forces of the chariot of progress. Among the most foresightful were Goethe, William Blake and Nietzsche. They saw what was coming. They warned us. But we did not heed their warning.

Nietzsche's announcement of the death of God was but a dramatic expression of the whole process of the destruction of traditional values and of the instrumentalization of all values. The climate was created in which it became very difficult for a rational person to justify intrinsic values, especially religious values. From the middle of the nineteenth century we witness the rise of a new set of values: first came utilitarianism; then relativism and nihilism.

MORAL RELATIVISM – THE CANCER THAT IS EATING US AWAY

Moral relativism is eating away the substance of the twentieth-century individual. We have made a virtue of moral relativism and of nihilism (of cynicism too), quite forgetting that they are undermining the very fabric of our life; not only of moral life, but life at large.

Moral relativism is a specific product of twentieth-century Western culture; almost an inevitable consequence of a culture undermined first by the mechanistic *Weltanschauung* of Newtonian science, then by seductive technology which, in a truly Faustian manner, almost persuaded us that we only live once and therefore we live recklessly.

Why is moral relativism almost an inevitable consequence of the scientific–technological world-view? Some of the arguments presented in chapters 2 and 6 are again relevant here. When the accepted dogma is that the intelligent person must be governed by scientific rationality alone, that religious beliefs and spiritual values are 'old hat', then moral relativism became not only an option but a necessity.

The centre of humankind, as rooted in religion, has been gradually destroyed by the combined influence of atheism and scientific rationality. At the beginning of the century cultural anthropology started to emphasize the *variety* of human cultures, and also emphasized a variety of values in these cultures. This cultural relativism went hand in hand with science's presumption that there are no absolute values of any sort. This cultural relativism indirectly created a nourishing ground for moral relativism.

Other phenomena of twentieth-century culture conspired to elevate moral relativism to an accepted and indeed cherished position. The disintegration of traditional conceptions of man, which brought about the existentialist concept of the individual, added to the feeling that we are all adrift, that *anything* goes.

Furthermore, the ethos of consumerism in a subtle and a pervasive way has been promoting the course of moral relativism. Let us notice that at the very heart of the consumerist philosophy (and of advertising which is its methodology, so to speak) lies the assumption that anybody can be persuaded to indulge and consume. Indulgence is a universal moral principle of advertising and of consumerism. This principle abhors all other principles, particularly of a deeper moral nature, which could make us immune to persuasion, and thus to continuous consumption. Let me emphasize this point: the pervading ideology of consumer society is that there is only one universal value – that of consumption. All other values are relative; we really don't need them, they might work as a restraint. Thus *the ideology of consumerism requires and demands moral relativism as a universal code of behaviour*.

We should make no mistake, moral relativism is an indulgence, a licence to do as you please, laxity, a lack of discipline and of responsibility. That is what a consumer society encourages in us, and indeed must cultivate in order to be effective. Therefore it should not come as a surprise to us that moral relativism meshes so well with all kinds of *me*-isms: hedonism, narcissism, epicurism. Each of them is a philosophy of pampering the ego. On the other hand, moral relativism does not go very well with altruism, and with forms of ethics that are *not* based on indulgence.

Culture is a very subtle and devious entity. It would appear that the scientific world-view, and the training of the mind of our young in the rigours of objectivity, have nothing to do with morality; and especially, nothing to do with the ethos of consumerism. But nevertheless, the connection is there. Scientific training – and this can be seen in *every* major university of the world – by its ruthless insistence on the value of facts, creates a value vacuum. By cultivating this value vacuum we invite moral relativism, indulgence and indifference, alienation and apathy, passing the buck and shrugging the shoulders, saying, 'Everybody does the same thing.'

Why should moral relativism be opposed? Because it has made life too cheap; both our individual lives and our societal life. Senseless killing of people without motive is a by-product of moral relativism. If *anything* goes, then why not to try the ultimate thrill – the taking of another person's life? When this happens in Dostoevsky's *Crime and Punishment*, as Raskolnikov takes the life of Alona Ivanovna, we witness a cosmic drama. In our times such acts are senseless, sordid and unspeakably morose affairs.

Moral relativism, though supported by so many intelligent and rational people, is a retreat and a form of cowardice; not really a moral position. Interestingly, it cannot be justified *morally*, for what can you say in its defence? That everybody does the same? This is not a moral justification and, besides, it is not true either. Like scepticism, moral relativism is self-defeating. If you take it seriously, if anything goes, then a *denial* of moral relativism goes as well. And if so, moral relativism as a moral proposition collapses. In the final analysis, moral relativism is an admission of an intellectual bankruptcy, and an inability to have a moral position. This may not be a palatable conclusion to the rationalist who so often *is* a moral relativist.

Moral relativism, though usually proclaimed as an innocent minimal moral doctrine, is not so innocent. As I have argued, it is either incoherent as a moral doctrine, and self-defeating in the same sense in which scepticism is; or, if it is upheld as a moral doctrine, with all the consequences following, it spells out a cryptofascist position, claiming that we are all different, therefore some are unequal, and those unequal have more right to satisfy their needs than others – particularly as there is no moral standard of equality and therefore no standard of justice. Beware of simple innocuous moral doctrines; they are a loaded dice. Beware of your own moral relativism; it is not such a nice, innocent posture as you might have thought. In addition to being a doctrine that leads to the cheapening of human life, moral relativism indirectly favours injustice: it favours the rich, the privileged and the manipulative, under the mask 'we are all different'.

THE LEGACY OF TWENTIETH-CENTURY MORAL PHILOSOPHY

Moral relativism has found an unexpected ally in twentieth-century moral philosophy. The main branches of traditional philosophy included ontology, epistemology, ethics, aesthetics and metaphysics. Ethics was traditionally that department of philosophy which dealt with morals. Analytical philosophy of the twentieth century decided that this was not good enough as ethics so often took a normative stance, urging us to accept a specific set of values; analytical philosophy has insisted that what is of utmost importance is the *meaning* of moral concepts. So ethics has been replaced by moral philosophy which dissects moral systems rather than offers any values of its own. For the last ninety years moral philosophy has been doing just this, quite forgetting that one cannot live by analysis alone.

No man has contributed more to the emergence of moral philosophy, in the sense we have just mentioned, than the British philosopher G. E. Moore (1873–1958). His epoch-making treatise of 1903, entitled *Principia Ethica*, was clearly an attempt to establish in ethics the sort of foundations that Russell and Whitehead established for mathematics and logic in their monumental *Principia Mathematica* (1903–10). Moore's approach and his techniques, of endlessly analysing moral terms, have had a profound and lasting influence on the entire scope of Western moral philosophy.

In pursuing the detached objective idiom of moral philosophy, which accepts no moral commitment but only analyses, we have helped to create a moral vacuum. The indirect result of replacing ethics with the analytical scrutiny of moral concepts is the relativist fallout. The historical heritage of human morality has been subjected to continuous analytical needles. Since normative ethics were declared to be passé, a mere remnant of pre-rational societies, moral relativism has naturally crept in as the lowest common denominator.

G. E. Moore was an astute thinker and very clever in identifying the various fallacies in the systems of others. Yet his whole philosophy is based on a fallacy (which I shall call Moore's Fallacy), and this was: confusing moral insight with linguistic insight. Analytical philosophy, since Moore, has been continually trapped by this fallacy. Those bright philosophers simply could not see that the whole programme had been based on a dreadful mistake – of confusing analytical clarity with moral insight.

A small child, or a very simple person can be told: do not kill. He may ask: why? We shall respond: because everything has a right to live. And he will understand. He will understand through his capacity for empathy; simply because he has got a *moral sense within*. When this moral sense is activated by a principle that touches the cords of our brotherhood with other beings, or illumines for us the idea of justice, then moral insight follows; and this is regardless of our analytical acumen.

There is the other side of the picture. A sophisticated analytical scrutiny of basic moral concepts by a brilliant logical mind may result in no moral insight whatsoever, particularly when the mind is insensitive, devoid of moral sense, eaten by scepticism and nihilism. Therefore these two things, the moral insight and the analytical insight, are quite apart. We have laboured under a tremendous illusion that linguistic insight is the key to moral insight. It is not! It suffices to notice that so many first-rate analytical minds, including moral philosophers, have demonstrated their exemplary moral insensitivity. One wonders whether this excessive analytical training – in dispassionate, objective analysis – has not contributed to the extinction of their moral sensitivity.

This is indeed a dark side of the alleged brightness of analytical philosophy. Analytical philosophers are unable to *see* in a deeper sense of the term. Moral philosophy does not understand that analytical understanding does not automatically yield moral understanding. These are two different acts. Morality is *sui generis*. Traditional societies have known this truth very well indeed. For this reason, while educating their young, they placed a special emphasis on developing the moral sensitivity.

We have a larger problem here, for analytical philosophy claims that all understanding is analytical understanding. This is its premise, which is demonstrably false. It is valid only when we *assume* a priori its validity. To attempt that which should be proven, namely, that all forms of comprehension are exhausted by analytical comprehension, is question-begging. In our practical life and in our spiritual life we know that there are a variety of forms of comprehension. Moral comprehension, aesthetic comprehension and biological knowledge, which manifests itself through a variety of sensitivities (see chapter 5), are all *sui generis*.

We must overcome Moore's Fallacy. We must recognize once again that rationality and ethics constitute distinctive domains of our experience; that being guided by the wisdom of traditional morality, as Hayek puts it 'may be superior to reason'. However, reason will not give up easily, and it is full of stratagems and sophistry. We have been numbed by it morally for a long time. It will take time and courage to reactivate our compassion and wisdom to see that moral insights and moral enlightenment cannot be derived from mere intellectual reasoning.

The whole mode of analysis of twentieth-century moral philosophers is a tribute paid to scientific reason. What we thus obtain is not *moral enlightenment* but *intellectual reasoning* about the structure of moral propositions. Analytical philosophy has mistaken rationality for morality.

In brief, philosophers have subconsciously attempted to make values comply with the structure of scientific reason. They have imposed the structure of scientific reason on moral discourse. The result is a set of stultifying abstractions. No wonder that people cannot recognize in the discourses of philosophers anything that has to do with moral and ethical dilemmas as they experience them.

ON THE IMPORTANCE OF FOUNDATIONS AND OF INTRINSIC VALUES

What kind of values do I uphold? What kind of universe do I live in? What kind of destiny do I pursue? These are the questions that are of importance to all of us. These are the questions that are at the heart of our existential malaise – whether we are intellectuals or ordinary workers. Because we have not resolved these questions in any way, we cannot find the key to harmony, we cannot resolve the problem of the meaningfulness of our individual lives; and on another level, we cannot find satisfactory solutions to large-scale environmental problems. They are all connected. When the overall harmony, cohesion and meaning evade us, we are at a loss as to how to deal with particular problems.

In brief, I will try to argue that the resolution of our environmental and ecological dilemmas lies in the matrix of our values. Unless we are able to see in depth what values we behold and how they control our behaviour, unless we are able to establish a new sound, sane and sustainable value-basis, all the dazzling expertise (based on a limited and fragmented vision), all the technological fixes, will be acts full of sound and fury, signifying nothing.

Ethics is not engineering. It does not ask the question *How to?* but *Why?* While developing ethics, we do not search for tools for fixing things. Instead we search for ultimate foundations which alone can justify our being in this universe. It is very important to bear this in mind in our instrumental age, when we are inclined to reduce everything to a technique. If ethics is a technique, then it is a technique of the soul, following quite a different route from that of present technologies.

We are impatient with general principles. We are impatient with philosophy. We want guides for action *now*. But that is the attitude of a technician. Ethics, on the other hand, tries to understand a deeper nature of things, and especially *why* should we behave in this way and not another way? Asking *why* questions sooner or later leads to foundations, and to foundation values. *Foundation values are a rock on which the whole ethical system rests – whatever its nature. If we do not accept some foundation values, nothing follows.* For foundation values give us the *raison d'être* for the whole system, its specific sub-values and its specific modes of action.

I call foundation values the first order values. I call the consequences of foundation values, the second order values. I call specific tactics and strategies for the implementation of the second order values, the third order values. Let us use some examples to illustrate the point:

We should work on legislators to pass appropriate bills to save environments. Our political action (to persuade the legislators) is in the realm of the third order values.

The justification for third order values is in the second order. Why should we work on legislators? Because we value environments. This is in the realm of the second order values. What is the justification of this one? A still deeper value which I call a foundation value. In the system of eco-ethics that I propose, this foundation value is rooted in the idea of the *sanctity of life*. The acceptance of the sanctity of life prompts us to protect other forms of life, prompts us to protect threatened habitats, as well as human environments in which life is in peril.

All conservation work, all environmental protection activities, are ultimately based on this deeper conviction of the sanctity of life. Let us be quite clear that if the premise of the sanctity of life is questioned or rejected, the whole design of conservation strategies, and all specific actions for saving environments, are left hanging in thin air. There is no reason then why we should engage in conservation strategies and ecological ethics.

This point has to be stressed over and again. We are so intoxicated with action that we often think that it is the only thing of value. But action has meaning only if it has a deeper meaning. *The meaning of action is determined by the deeper principle this action serves.* (See once again the diagram on p. 12)

Foundation values (in any system of ethics) are the terminal points. They provide the justification for the meaning of a given ethics, as well as the justification for the actions performed under the auspices of this ethics. If we deny foundations, nothing follows.

The important ethical systems of our times are those which clearly spelled out their foundation values and built on those values.

Thus for Gandhi, the foundation value was *Ahimsa* – non-violence.

For Schweitzer, it was reverence for life.

For Aldo Leopold, it was the sacredness of the land.

The eco-ethics that I have proposed is based on the idea of the sanctity of life. From this idea, follows the ethical imperative of reverence for life, which is another formulation of the idea of the sanctity of life.[2]

Ultimate ethical principles underlie and justify our rational strategies and practical choices. Such has been the story of great ethical systems of humanity. We shall do well by following the wisdom of past ethical systems without necessarily embracing their specific principles. In this context we have to see clearly that relativism of values does not represent a value position but represents an abdication from holding a value position.

To postulate intrinsic values does not mean to postulate either absolute values or objective values, but values that bind us together as a species endowed with certain attributes, propensities and common imperatives. Because of *analytical* difficulties in justifying intrinsic values, the notion has been abandoned by many philosophers. This I consider a mistake. We need intrinsic values as the backbone of our ethical reconstruction and of conservation strategies. I will attempt to argue that eco-ethics are a set of new intrinsic values. But first let me attempt to clarify the status of intrinsic values. In this endeavour I will build on the proposition that the intellectual insight and the moral insight are two distinctive entities and therefore we must not attempt to subsume the moral insight under the intellectual one because then we wipe out from ethical values what is *sui generis* in them.

My basic proposition is that it is our value consciousness or the *axiological consciousness* that is of primary importance in value-recognition. It is this consciousness that informs and guides our intellectual consciousness regarding values. This insight is of great importance, for it leads to a new clarification of intrinsic value.

There are no intrinsic values beyond our consciousness as a species and independent of it. Things are not valuable in themselves, it is our consciousness that makes them valuable. What I propose is not an expression of subjectivism. Our intrinsic values are species-specific. In this sense they are inter-subjective.

Some philosophers have mistakenly argued that because values that are species-specific are neither objective (in the ontological sense) nor absolute, they must be subjective. To argue this is to misunderstand the meaning of the term 'subjective'. Subjective is that which is limited to an individual human subject. On the other hand, that which is species-specific is certainly trans-subjective, ergo inter-subjective. It is in this sense that ecological values are trans-subjective. I will further argue that it is in this sense that ecological values are both inter-subjective and intrinsic. But intrinsic values need not be objective – unless you are a Platonist.[3]

Thus between the Scylla of subjectivism and Charybdis of objectivism there lies an inter-subjective justification of intrinsic values as assessed by our axiological consciousness which is species--specific, therefore trans-subjective. If moral values arise in our axiological consciousness, only in this consciousness can they be justified. If so, then it is inappropriate if not indeed fallacious to attempt to justify them with reference to objects outside our consciousness, that is to say, by insisting that moral good must reside in objects in the external reality. *Objects and situations by themselves are neither good nor bad. It is our value consciousness that makes them bad or good.* Hence, no attribution of intrinsic value as ontological claims independently of our valuing as a species. Once again: ontology has to do with the description of things; *axiology* has to do with our *value* consciousness. Do not mistake axiology for ontology.

Thus we can uphold intrinsic values and justify them without slipping into Platonism or subjectivism or relativism. Yet a new justification of intrinsic values requires a new moral insight. This new moral insight for me, and many others in our times, is the recognition that nature is not an object to be trampled upon and that all other beings were not created for our use, but that nature is alive and we are a part of it; and that all other beings are our brothers and sisters in creation. This insight leads (on the ethical level) to the annunciation of the principle of the sanctity of life or reverence for life – from which an ecological ethics is derived.

REVERENCE FOR LIFE AND OTHER ECOLOGICAL VALUES

Every new ethical insight spells out a new perception, which usually leads to an articulation of a new relationship between us and our cosmos. These new perceptions spell out, or at least indicate, our new responsibilities and new obligations, which sometimes are formulated as commandments. An example of a new ethical insight is Leopold's land ethic:

> A land ethic then reflects the existence of an ecological conscience, and this in turn reflects a conviction of individual responsibility for the health of the land. Health is the capacity for self-renewal. Conservation is our effort to understand and preserve this capacity.[4]

Leopold is one of the champions of ecological awareness. Leopold's perceptions clearly spell out new values which bind us to the land. The responsibility for the health of the land is one of the essential obligations that we undertake, living on the land, with the land, off the land.

For Leopold, our responsibility for the land is an *obligation* which does not need any further justification: we are it; it is us. It is a good thing *in itself* to take care of the land. We must do it because it is our responsibility, irrespective of the yields. Eco-values developed in this book are an extension and continuation of two insights: of Leopold's land ethic and of Schweitzer's reverence for life.

Among the intrinsic values for our times, the most important is reverence for life – born of a vision and conviction of the sanctity of all living things. This vision is actually easy to accept; that is, before we become influenced and really corrupted by scientific thinking or, to be more precise, by *mechanistic* thinking. This vision is entertained and accepted in the world-view of American Indians. This vision is natural to small children in our own civilization. We have to relearn to appreciate the beauty of this vision.

Yet the scientifically rational mind finds it hard to entertain anything that is sacred. The very term *reverence* is difficult to accommodate within the rational frame of discourse. Even such an innocent concept as 'vision' makes some feel uncomfortable. We are told that we do not live in the world of visions but in the world of harsh realities. But we are told wrongly. The mechanistic conception of the universe (and the 'harsh' realities following it) is no less a vision than the reverential conception of the universe. Nature can be looked at in so many different ways. We invent our metaphors, and then find in nature what these metaphors assume. This is what our cosmologies are about.

Responsibility is another intrinsic value of eco-ethics. You cannot exercise reverence without responsibility; ultimately responsibility becomes reverence. Responsibility is part of the meaning of reverence. The two co-define each other. There is a whole negative historical connotation attached to the notion of responsibility within the protestant ethics. This negativism is a baggage which we must throw out, so that we can see the concept in its true light: as a radiant principle which enables us to revere the world and appreciate its transphysical dimensions. Responsibility is an ethical principle in the sense that if we understand the unity of life, and the fact that we are a part of it, and one with it, then we must take responsibility for life, for all life; there is no other way. *Thus the right understanding of the world and in particular an understanding of the sanctity of life implies our responsibility for it.* It is really that simple. Responsibility is the connecting link between ethics and rationality. Rationality without responsibility is monstrous, as it has been shown by the example of the German Nazis. Ethics without responsibility is empty – as has been shown to be the case with formal ethical systems. Responsibility is the spiritual bridge which makes of rationality a human rationality, and of ethics a nourishing river of the meaning of our lives.

The larger the scope of responsibility we assume, the larger we become as people. If we assume no responsibility, we are hardly human. Escape from responsibility, which the indulgent society perpetuates, is an escape from our own humanity. If we want to shun *all* responsibility, there are only two ways: live as a complete hermit, away from everybody (although this is not really possible, as there will always be the company of birds, plants, the mother earth, and the father sun); or – more radically – commit suicide, and this is the last thing that a responsible person should do. The truly great lives, like Gandhi's and Mother Teresa's, are lives pregnant with immense responsibilities.[5]

Another intrinsic value of eco-ethics is *frugality*. Frugality must be handled carefully so that it is not confused with abnegation or destitution. Frugality is an altogether positive value, a form of richness, not of poverty. Frugality is a vehicle of responsibility, a mode of being that makes responsibility possible and tangible, in the world in which we recognize natural constraints and the symbiotic relationships of a connected system of life. To understand the right of others to live is to limit our unnecessary wants. The motto in one of the Franciscan retreat houses reads: 'Anything we have that is more than we need is stolen from those who have less than they need.' Is this too strongly expressed? People in the Third World countries would not think so.

On yet another level, frugality is a precondition of inner beauty. We are frugal not just for the sake of others but for our own sake. *Frugality is an optimal mode of living* vis-à-vis *other beings*. A true awareness of frugality and its right enactment is born out of the conviction that things of the greatest value are free: friendship, love, inner joy, the freedom to develop within. Indeed, if you want to buy these things, the more you are prepared to pay the more surely you will destroy their inherent worth.

On a higher level still, *frugality is grace without waste*. The greatest works of art are frugal in this sense. Grace shines through them, and it is grace without waste. Thus we must cultivate meaningful and elegant frugality. To do so, we will need to develop a new language so that we are not constrained by past connotations which attribute to frugality dreary and tiresome characteristics. Let us therefore be supremely conscious that frugality is not a prohibition, not a negative commandment (be frugal or be doomed), but a positive precept: *be frugal and shine with health and grace*. You cannot live in grace when you live in poverty. On the other hand, you cannot live in grace when you wallow in spurious luxury. Grace is the divine middle. Aristotle was already aware of the beauty of frugality when he wrote that the rich are not only the ones who own much but also the ones who need little.

The true image of frugality is Gandhi, whose life was slender in means but incomparably rich in ends, woven into the tapestry of others, confirming the unity of all and affirming justice for all. An impossible dream? But life is made of the stuff of dreams.

Frugality is an aspect of reverence. You cannot be truly reverential towards life unless you are frugal, in this present world of ours in which the balances are so delicate and so easy to strain. The three basic ecological values, reverence, responsibility and frugality, are so interwoven with each other and interconnected that the meaning of each presupposes the existence of the other two. In brief, the precepts – be reverential, be frugal, be responsible – are ethical commands following from our deeper insight of the connectedness of life, of the unity of life, and of its essential fragility.

For fear that the importance of frugality may still not be properly appreciated, I would suggest that we now change the slogan of the French Revolution to read:

Humanité, Fraternité, Frugalité

Among specifically ecological values another should be singled out and this is *diversity*: thou shalt act in such a way as to increase and preserve diversity, and diminish or stop any trend towards homogeneity. Diversity may be seen as a questionable candidate for a moral category. Usually we take it to be a descriptive term. In biology and botany it has to do with the richness of habitats. But this very richness at a certain point may be perceived as a moral attribute: diversity = richness = complexity = life. Thus if we want to preserve and maintain a thriving life, whether of an ecological habitat, of a culture, or of an individual, diversity is a *sine qua non*. As such, it is our *moral responsibility* to maintain and increase diversity. This is well understood in terms of genetic diversity. We now keep a pool of genetically diverse seeds to preserve the genetic richness of life. To keep and maintain this pool is good in itself; it is keeping life alive. This is the purpose of all ethics – to keep life alive, to help it and above all to enhance it. Ontological, cultural and genetic diversities are prime movers of life's richness.

Morality is a peculiar attribute of human life that is alive. It is therefore a moral imperative to maintain those forces that make life alive, that bring life to fruition and abundance. When Jesus says, 'I am life abundant,' he utters a moral statement.

Eco-justice is another value specific to ecological ethics. Eco-justice means justice for all. Justice is a venerable, ancient concept. We are all familiar with it; particularly as it applies to us. We all demand justice, especially for ourselves. Nearly all past ethical codes accept justice as an integral part of moral behaviour. Yet in traditional moral codes justice is limited to the human universe. Sometimes it is limited to a particular religion. Then 'infidels' can be mistreated with the justification, 'They are not the children of *our* God.'

Eco-justice as justice for all is simply a consequence of our ecological reverence; it is also a consequence of the idea of responsibility for all and of the perception of the interconnectedness of all. If the cosmic web embraces us all, if it is woven of the strands of which we are parts, then justice to the cosmic web means justice to all its elements; to all brothers and sisters of creation, as the American Indians would say.

It is difficult to render justice to all in this complex and contingent world of ours. This we know. But it is our moral duty to attempt to do so. Moral principles can be annunciated even if it is difficult to live by them.

Ecological values are exactly like other traditional values – they are ideal signposts and imperatives for action. The fact that they may be difficult to implement in practice in no way negates their importance and desirability. Great value-systems of the past were established not because they were *easy* to practise but because they expressed the imperatives which were deemed important to safeguard life, meaning and human dignity. All genuine values serve life and the quest for human dignity, and so do ecological values.

The ecological values here discussed are only a core. Their application may differ in specific circumstances and they will need a creative extension in various walks of life, just as traditional values did. Let me emphasize that what is proposed here is not a set of specific strategies but general principles, an underlying matrix of values to guide our thinking, perception and action in our attempt to establish right relationships with the earth, with nature and with other cultures.

ECOLOGICAL VALUES AND SUSTAINABLE DEVELOPMENT

Conservation is a mode of thought and action through which we express our responsibility to nature. Conservation is a very special kind of activity. It expresses care, piety, love, attention and a lot of hard work to save what we deem worth preservation.

In one sense conservation may be considered an ethical act in itself: an act of caring to the point of fighting for what you consider important to preserve, as I have argued elsewhere.[6] On a deeper level of analysis, conservation is a set of strategies for implementing ecological values in the primary sense. We design various tactics and stratagems to save what is worthy and necessary of protection – be it our individual life, our family life, our social and civic life, or eco-habitats – because we value life as such, because we uphold the principle of the sanctity of life. Strategies come after the primary values are established and accepted. We should clearly see the centrality of reverence for life as the basis for sustainable development. To develop or not to develop is not the question, but rather *how* to develop. The central question is: what are the aims, goals, purposes and ends of development?

The ultimate end of all development is life. Development serves life and positively contributes to life. If life is taken out of the equation, all notions of development become meaningless. Furthermore, it is not merely life that we value, but quality living – for all creatures. *What is at the kernel of development is not just biological life, but life of meaning, dignity and fulfilment.* Unless we respect this conception of life, we need not bother about development.

Thus, when carefully examined and unpacked, the idea of development presupposes a life of meaning, dignity, fulfilment and self-actualization. If we look perceptively, we shall see that these concepts of development, which are truly comprehensive, assume that development serves a variety of life, not just economic ends.

We may approach the issue from the other end: if we accept the notion of reverence and of the sanctity of life, this enables us to see immediately that development is not only an economic phenomenon but also that it is a vehicle for the amelioration of human life on all its levels. Reverence for life does not deny the importance of the economic factor; the satisfaction of basic wants is part of the life of dignity.

At this point we would wish to propose a new concept, that of *reverential development*. Reverential is not merely a nice adjective which we attach to traditional (economy-determined) development, but is of the essence if development is not to become a bulldozer crushing frail balances of nature and leaving in its wake stupefied consumers. It would thus appear that reverence for life and development are intricately connected in the framework of thinking and action in which the meaning of human life prevails, and in which respect for nature is part of our conscious and compassionate interaction with all there is.

Ecological ethics, as based on reverence for life, is universal in the sense that in all cultures and major religions there is a latent premise of the worth of life, and indeed of its sanctity. In most traditional religions this sanctity derives from God. But this sanctity can be derived from Evolution itself, as I have argued in this book.

The fact that reverence for life (which we advocate here) coincides with so many religious ethics, based on the sanctity of life, should only reassure us that there is an underlying core of ethical values common to all people and most religions. *Ecological ethics represents a new articulation of traditional intrinsic values.* It represents the search for meaning, dignity, health and sanity at a time when the planet is seriously threatened by inappropriate development.

In proposing a new form of development – reverential development, based on ecological values – we wish simultaneously to bring about sustainability to the planet, dignity to various (exploited) people and unity to a human race fractured by inappropriate development. Thus reverential development is unitary in the broadest and deepest sense:

- It combines the economic with the ethical and the reverential.
- It combines contemporary ethical imperatives with traditional ethical codes.
- It attempts to serve all the people of all cultures.
- Last but not least, it promises to bring about a truce between humankind and nature, including all its beings.

A SYNTHESIS OF VALUES

Let me now contrast the three basic trees of values: the traditional religious, the modern secular, and the emerging ecological. Once we table them together, we immediately see how different they are, and also that they demand of us different forms of behaviour, indeed, delineate different structures of our being.

The table displays the three sets of values in relation to each

Technological Values	**Ecological Values**	**Religious Values**
mastery	reverence	submission
control	responsibility	worship
power over things	frugality	grace
homogeneity	diversity	obedience
individual justice	eco-justice (justice for all)	God's justice

other. As it shows, the dominant value in a religious society is that of submission, in a scientific-technological society it is that of mastery and in the emergent ecological society it is that of reverence.

If we take one penetrating look at the table, we immediately notice that the values of the technological culture – mastery, control, power, linearity and individual justice – are the values (revealingly!) that *regulate our relationships with objects*. It is very curious indeed that this perception has somehow evaded us: that not only do we deal with different values but that the entire focus has been shifted from our relationships with God and our fellow humans, to our relationships with objects. This alone was bound to impoverish us in more senses than one.

Though startling as this fact is, yet in another sense it is not surprising at all. The secular world-view and secular values were *meant* to challenge and replace older religious values. The irony is that secular values were meant to add a cubit to our stature, not reduce us to the stature of objects! The whole enthralling dialectic, often murky and bloody, has played itself out. We see the pendulum swing from one extreme to another: from religious values, spelling out our submission to God, to secular values, spelling out our mastery over everything else.

Ecological values signify an act of creative synthesis. For this reason they are squarely in the middle, conjoining the past traditions, and transcending them. Nietzsche, Blake or Goethe could not have embarked on a new synthesis for the scientific world-view was still in its unfolding swing. Now the scientific world-view is receding. It lies exhausted and barren – crying for help. We are witnessing the dawn of the ecological epoch.

Life is a process of continuous synthesis. We cannot live in the strait-jacket of conflicting value-structures that tear us apart, or at any rate continually confuse us with their incompatible demands. A synthesis of values is a *sine qua non* for our wholeness and for our peace of mind.

So far I have said nothing about *love*, which shines supreme among the most important ecological values. Love is the most profound, the ultimate form of conservation in action; the supreme ethical act. We have lost the power of love; to regain it, we must first develop a sense of reverence, which is a precondition of love.

Great ethical and moral traditions are usually related to one single overarching concept. In Buddhism it is compassion; in Christianity it is love; in ecological ethics it is reverence, which does not quarrel with the other two, but actually sets a plateau from which the road to the other two is clear and unobstructed.

CONCLUSION

The genetic code makes us into distinctive biological beings, but after the genes have performed their task, we are still only half finished; only after we acquire culture and values, only then are we complete as human beings.

The values which traditionally clarified our status as human beings and which articulated our moral conscience, as well as made us sensitive to culture, were inspired and guided by religion. Religious values tended to be absolute as they were rooted in the absolute ground of being – God. With the twilight of religion, religious values have become gradually less and less significant, particularly as the absolute ground for their existence, God, has been questioned, challenged and often abolished – 'God is dead,' said Nietzsche. The triumphant march of the secular world-view signified the rise of scientific-technological values, which increasingly tended to be utilitarian, instrumental, relativistic, freeing us from old absolutes but also cutting us from the sources of our spiritual sustenance and deeper meaning.

Moral philosophy of the twentieth century, stemming from G. E. Moore's *Principia Ethica*, has been more interested in the *meaning* of moral concepts than in understanding the *nature* of moral insight, the particular faculty that makes us moral agents. As a result, we have accumulated an impressive body of linguistic analyses that have not led to any significant moral enlightenment. On the contrary, these semantic analyses led to subjectivism and relativism. By mistaking the intellectual insight for the moral insight, we have inadvertently undercut all inter-subjective codes of values. Yet in order to be held together as a human family, we must have the glue that sticks us together. In brief, we need new values, new normative ethics to live by, not sharp analytical instruments for analysing past values.

Although we have claimed in the West to be liberated and enlightened, we have been trembling in our boots when confronted with the dictates of scientific rationality, which has been a censor, a tyrant and an absolute ruler of what counts as a valid discourse. We have uncritically followed the gospel of instrumental (scientific) rationality, which has reinforced the overall drift towards relativism and nihilism.

Every act of perception, every meaningful interaction in the real world, is guided or inspired by values. *Unexamined values lead to moral nihilism.* Moral nihilism, like moral relativism, is not free of values: it tacitly promotes and perpetuates decadence and escapism, and not infrequently justifies exploitation. The satisfaction of the ego is the only intrinsic value that moral relativism acknowledges; a far cry from a system of ethics which would be sustainable in the long run.

Ecological values bind us to nature, to the earth and to each other. They are our new intrinsic values. They are not absolute; but they are not subjective or relativistic. They are species-specific, and thus inter-subjective. They bind us together, and nourish us together, because we are a species of a certain kind connected with life at large in specific ways. All value-systems are ultimately justified by life. Spelling out what is life-enhancing in the long run, and what at the same time assures the optimum harmony among the species, is tantamount to spelling out a set of inter-subjective values. In this sense, ecological values are intrinsic values for our times. Most important among these values are: *reverence* for life, *responsibility* for all there is, *frugality* in our lifestyles, and *justice* for all. Environmental ethics is not enough, for too often it becomes an instrumental ethics, based on a cost-benefit analysis. A genuine ecological ethics must be based on some intrinsic values, which are species-specific.

To protect life is morally right. Conservation and sustainable development are based on the premise that we must preserve the heritage of the earth and all its species, and also that we must be mindful of future generations. The very idea of 'development' is not an objective or merely descriptive category; 'development' is a normative term, loaded with values. There is no logical necessity for any development to occur. We cherish meaning, fulfilment and the idea of living to our full potential. To these ends genuine development (which I call reverential development) should be directed. The right concept of development must be based on wisdom. Otherwise it becomes an exploitative development which benefits some (in the short run) and ruins most (in the long run). Pursuing wisdom in the long run means enacting right values.

Wisdom for our times I call ecological wisdom; values that express ecological wisdom I call ecological values. Strategies and tactics that truly bring about sustainable development that benefits all, or at least most, are based on ecological wisdom, and thus on ecological values. The pursuit of reverential development is therefore identical with the pursuit of ecological values and ecological wisdom.

9

Ecological Consciousness as the Next Stage of Evolution

FROM RELIGIOUS CONSCIOUSNESS TO TECHNOLOGICAL CONSCIOUSNESS

Some six hundred years ago, religious consciousness was dominant in the Western world. This consciousness was wedded to Christian cosmology. As we shall see later, every form of consciousness is connected with a form of cosmology, by which it is engendered and which it articulates. Religious consciousness and Christian cosmology went hand in hand with each other, co-defining and mutually supportive.

Guided by religious consciousness, people were adjusted to a universe that was overlooked and steered by God. Their daily affairs were regulated by the awareness of the omnipresence of God, and by the presence of his representatives – the clergy. The point to be emphasized is that the entire field of consciousness of the medieval person was shaped and pervaded by the images of God, by the idea of responsibility to God, and by the desire to be saved and redeemed in God's heaven. Then things began to change. After the messy, turbulent and effervescent Renaissance, a new epoch gathered momentum in the seventeenth century. The medieval Christian cosmology was questioned. In the process, religious consciousness was undermined and punctured in many places.

Secularism emerged as a new umbrella under which a new consciousness was being crystallized, in marked opposition to the earlier religious consciousness. This secular consciousness would be articulated in time in the form of technological consciousness, but for the time being (throughout the seventeenth century), this consciousness was groping for articulation, for a new distinctive shape.

The Humanist movement of the Renaissance was the first step towards the new non-religious consciousness. Renaissance humanists repeated after Protagoras, 'Man is the measure of all things.'

The next crucial step in the rise of technological consciousness was the quantification of the cosmos. Only with the advent of the mechanistic world-view did the situation change dramatically. The combined efforts of Bacon, Galileo and Descartes (and scores of others, of course), resulted in a new cosmological matrix, a new world-view which conceived of the universe as a clock-like mechanism. From this point on, the developments were rapid, far-reaching and with staggering consequences.

The new cosmological matrix requires that all phenomena, to be recognized as valid, must be physical in nature or reducible to physical components. The important relationships to do with these physical phenomena should be expressed in quantitative laws. Thus, the physical and the quantitative are enshrined. Over time this led to some significant consequences, such as:

- the worship of objective knowledge;
- the increasing quantification of all phenomena;
- the narrowing of the focus of our vision and our inquiry; and last but not least
- the elimination of the sacred.

A further consequence was the growing process of alienation – a direct result of the process of atomization and quantification. As we split everything into separate atoms, larger wholes were disintegrated. We no longer had a sense of wholeness, but rather a sense of isolation, separation, detachment, in brief – alienation. Psychological alienation was the result of conceptual alienation.

Another consequence of the mechanistic approach to the cosmos was the growing worship of physical power, indeed, a sense of intoxication with power, and an obsession with it. A corruption of power, to signify the coercive physical power, was a consequence of mechanistic cosmology, which – with an extraordinary consistency – eulogizes the physical, the quantitative, the manipulative, the controlling (see discussion in chapter 6).

It is at this stage that mechanistic cosmology became absorbed by the entire Western culture. It has come to dominate minds, leading to the emergence of a distinctive form of consciousness, which I call technological consciousness.

We must stress with due gravity that technology is not to be thought of as a chest of tools – indifferent in themselves, bad or good only according to our use. This is a rather atomistic and naïve view of technology, which technology itself attempts to perpetuate. At this stage of history, technology is so all-pervading that it is a form of consciousness. When we think technology we think 'control and manipulation'. *Technology is a vision of reality, not the use of tools*.

When we interact with the world via technology, we never think how to be benign and compassionate and loving, but always how to be efficient, controlling, assertive. This attitude of controlling and manipulating is now a part of the mental make-up of Western people.

Another intriguing, or shall I say fascinating, aspect of technology is that although technological consciousness is supposed to be secular through and through, it contains its own transcendental programme, its own form of divinity. It seeks the divinity of the human here on earth – by liberating us from old yokes, giving us dignity and freedom, and by creating us in the image of the industrious god who can do everything for himself.

Now, to seek fulfilment and realization on earth, through our own effort, is an admirable project. The trouble begins when we amass too much power with which we destroy natural habitats and by which we become so intoxicated that we forget our place on this planet. In the absence of higher values and some form of wisdom concerning human destiny, the amassing of power is a very dangerous thing, leading as it does to unbridled arrogance and ultimately to hubris.

Let me now summarize succinctly the characteristics of

technological consciousness. When we view its overall structure, its overall mode of operation, technological consciousness reveals itself as:

- Objectifying
- Atomizing
- Alienating
- Power dominating
- De-sacralizing
- Geared to the eschatology of consumption

This last point needs to be elaborated upon. What is the eschatology of consumption? Eschatology is the subject that concerns itself with the ultimate ends and goals of human life. In the absence of far-reaching transcendental goals, while religious and spiritual values have collapsed, consumption has become an imperative of our life, an overall goal, a form of fulfilment, the focus of aspirations.

In its clumsy and indirect way, consumption has become a form of salvation, thus an eschatology. Consuming new toys, new cornflakes, new cars, new televisions, new make-ups, new computers, is not dangerous in itself. What is dangerous, and unhealthy to our psychic lives, is when this process of consumption becomes a kind of religious urge promising happiness, fulfilment, salvation. At this point technology has become a form of eschatology.

In brief, after we have emptied the universe of the sacred, of the spiritual, of intrinsic values, after we have declared the physical, the objective, the coldly rational as our new deities, and after the traditional cultural patterns have been disintegrated in the miasma of atomistic thinking, what has emerged as our new eschatology is the drunken pursuit of power and of stupefying consumption.

What has also emerged is a new image of the human – the Faustian who celebrates his day by seeking gratification here and now. The Faustian maintains that one only lives once, therefore one lives dangerously, at whatever and whosoever's expense, even if it means the ruin of future generations and the destruction of ecological habitats. The Faustian is the human manifestation of rapacious technology, a symbolic acknowledgement of the ascent of naked power and the simultaneous waning of human spirituality.

While analysing the fallout of technological consciousness, I am not forgetting the other side of the coin, namely that technology has been a noble dream that has turned out sour, that the mechanistic world-view was once a ladder to freedom from the constraints and oppressions of religious consciousness. Nor do I wish to question the obviously beneficial aspects of technology – comfort, the rise of the material standard of living, the sense of freedom of movement (if only illusory), the elimination of contagious diseases.

Yet as we approach the twenty-first century, technological consciousness appears increasingly as a menace with its ruthless efficiency, its uncontrolled increase of power and its lack of any compassionate accounting. Technological consciousness simply does not add up. It produces too many diseases of its own, and the price for comfort and other amenities is simply not worth paying. This is the realization which we, as a society, have come to only gradually.

Technology is a story of success which has been so stupendous that it has become our nightmare. The pendulum has swung too far in the direction away from religious consciousness. Thus, we should seek to re-establish a new balance, and in the process attempt to balance our own lives.

Ecological consciousness, outlined in this chapter, is a synthesis. Religious consciousness was the thesis. Technological consciousness was the antithesis. Ecological consciousness is the synthesis as it marks a return to the spiritual without submitting to religious orthodoxies and religious dogma; and as it seeks social amelioration and justice for all without worshipping physical power and without celebrating the aggressive nature of the human person.

I am not claiming that ecological consciousness has arrived, is fully articulated and comfortably dwells in us. Rather, I am suggesting that in reaching out, we bring to fruition that which is dormant and which wants to be awakened and articulated. Our projection is part of our articulation and is an essential part of the creation of ecological consciousness. As we dream so reality becomes. The arrival of ecological consciousness is augured by many signs. Recently a powerful voice in its favour was raised in the Vatican. On the Day of Peace, 1 January 1990, Pope John Paul II issued a pastoral letter, which is the first ecological letter in its entirety, and which proclaims that 'ecological consciousness is emerging. It should not be suppressed but on the contrary encouraged and cultivated, so that it finds its expression in concrete programmes and initiatives.'

THE RISE OF ECOLOGICAL CONSCIOUSNESS

All modes of consciousness are rooted in history and have a historical character. They arrive at a certain point in history and then disappear, or are profoundly modified, at another point in history. Thus consciousness is determined by history. But, in its turn, it determines history. In our times we have been witnessing the *greening of consciousness*. Simultaneously, another, less easily perceptible, process is taking place: the greening of the world religions. Let us be aware that from the time of the Congress of Assisi (1986), where five major religions were represented, the ecological interpretation of these religions has acquired an important momentum.

The forerunners of ecological consciousness were the ecology movement, on the one hand, and various schools of humanistic psychology, on the other. In their respective ways they were against the temper of the mechanistic age. Both have emphasized holism and the irreducibility of large complex wholes to their underpinning components, ecological habits and human persons. Both these movements were a challenge thrown to the rationality of the mechanistic system. Both movements professed a new type of holistic rationality.

Moreover, in a certain sense both those movements possessed a religious flavour. They offered not only new intellectual vistas but a form of liberation. This liberation, although not always explicit, was meant to give us freedom from the deterministic and mechanistic shackles. Holism, which both movements emphasized, was the first step to liberation. Subconsciously we have grappled towards a new religion.

What are the chief characteristics of ecological consciousness? I shall enumerate six such characteristics and contrast them with the respective ones of technological consciousness. I do not claim that these six characteristics completely define the scope and nature of ecological consciousness, but we have to simplify. To simplify is to understand.

Ecological consciousness	vs	Technological consciousness
holistic	———	atomistic
qualitative	———	quantitative
spiritual	———	secular
reverential	———	objective
evolutionary	———	mechanistic
participatory	———	alienating

A more appropriate form of expressing the nature of ecological consciousness would be through a mandala, as each of its characteristics feeds into and feeds on each other, co-defining each other.

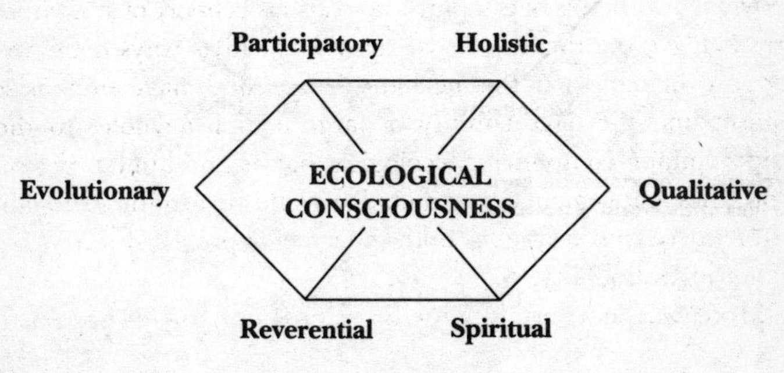

We could represent both mechanistic consciousness and ecological consciousness in one dynamic, open-ended model (see diagram below), which immediately shows how ecological

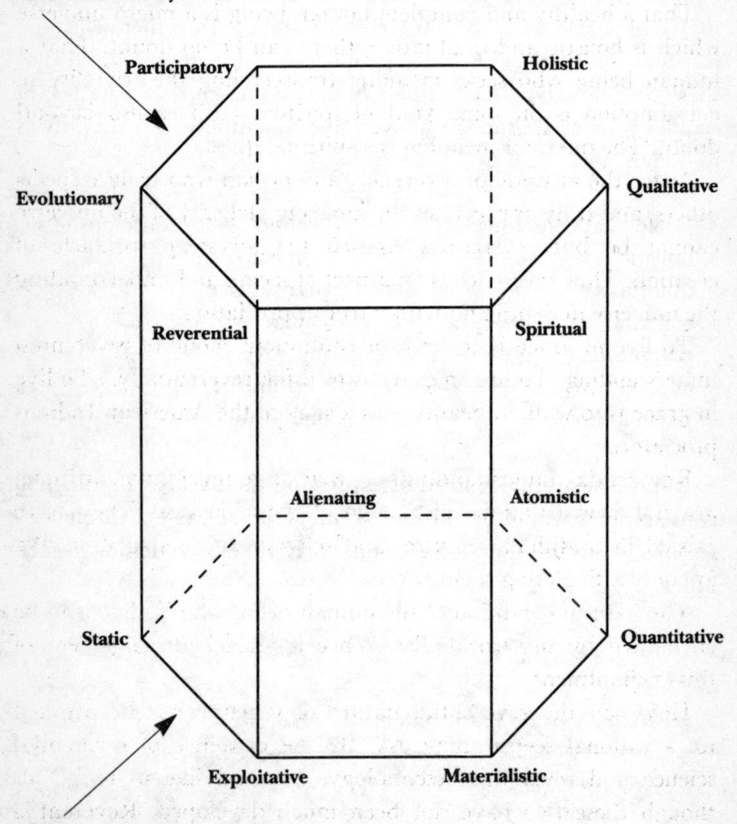

A model of consciousness

consciousness grows out of mechanistic consciousness and at the same time positively transcends it; and which also shows that both mechanistic and ecological consciousness are historical entities. The model thus demonstrates that we can and shall develop forms of consciousness beyond the ecological form.

Let us now discuss these characteristics and see what it all means for our own lives and for our perception of the universe.

That a healthy and complete human being is a micro-universe which is holistic and qualitative, there can be no doubt. That a human being who seeks meaning transcending the triviality of consumption is on some kind of spiritual path is also beyond doubt. The quest for meaning is a spiritual quest.

As for the attitude of reverence, any person who truly respects others, and truly appreciates the amazing alchemy of the universe cannot be but reverential *vis-à-vis* the awesome spectacle of creation. Thus reverence is an aspect of seeing and understanding the universe in depth and with a true appreciation.

To live in grace is to be in a continuous mode of reverential understanding. To live in grace is to think reverentially.[1] To live in grace is to walk in beauty – as a song of the American Indians proclaims.

Reverential understanding, as well as a reverential attitude, are not new creations of ecological consciousness. They have existed in traditional cultures and religions for a long time. We are only articulating them *de novo*.

The natural condition of the human being who is alive is to be enchanted by the world. Reverence is an acknowledgement of this enchantment.

How can the reverential nature of the universe be attested to, a rational sceptic may ask. By the existence of reverential science and reverential technology, I would like to reply, although these two have not been much developed. Reverential technology is something that the West needs to develop, as a fusion of existing technologies with a deeply compassionate attitude towards all creation. The various systems of yoga represent a form of reverential technology, and we should not raise an eyebrow or dismiss the idea simply because we are used to thinking about technology as physical tools engaging us with the physical world. Our interactions with the universe are numerous and subtle. Any tool or technique that engages us with the universe is a form of technology, and yoga systems are the techniques of the soul.

THE REVERENTIAL UNIVERSE

Let us once again reflect on the nature of cosmology. Cosmology makes assumptions about the universe *in toto*. And then, acting on these assumptions, it finds in the universe what it has assumed to be the case. Such has been the story of most historically known cosmologies. Cosmologies *do not prove* the existence of this or that attribute which they assume about the cosmos. They proceed as if this attribute was inherent in the structure of the universe, and then build large patterns of perception and knowledge that vindicate the assumed existence of a given attribute.

Let me underline an important point: *nothing reveals itself in the cosmos unless we assume it to be the case*. If we don't assume that the universe is physical in nature, we shall never be able to elicit this attribute from the universe. We have to start somewhere. Cosmology is a game of assumptions. These assumptions are not proven, they are made. And then acted upon.

This process of assumption-making or of creating cosmologies is a pre-scientific process, both in the historical and the epistemological senses. Science has little to say about this process because this process precedes science, and is therefore outside the jurisdiction of science. Western science comes into being only when a cosmology of a certain kind is assumed, namely mechanistic cosmology. Therefore, within the sphere of general cosmology science cannot be the arbiter of the validity of other cosmologies because it is in the service of the cosmology from which it came, namely the mechanistic one.

Thus, we need not worry about the verdict of science while we involve ourselves in the creation of non-mechanistic cosmologies. If science attempts to interfere with our new cosmological designs, we can tell science to go to hell – back to the mechanistic cosmology to which it belongs. For in trying to interfere with other cosmologies, mechanistic science oversteps its own domain and competence.

Returning to the point concerning the reverential nature of the universe: we need not be anti-scientific or ignore the existence of science for that matter; we need to observe, however, that there is nothing in the structure and the language of science that forbids us to view the universe reverentially.

Cosmologies are a matter of will and of vision. If we develop a reverential attitude towards the universe, if we articulate these forms of thinking, perception and behaviour that enable us to walk in beauty, we shall dwell in the reverential universe. The universe will be reverential because we will have made it so. Thus the universe *is* reverential if we have the capacity to interact with it reverentially.

To the divine mind the cosmos is divine. To the crass mind the cosmos is crass. To the monkey's mind the cosmos is monkey-like. These propositions must be taken with all seriousness. For it is the mind that rules over the unruly cosmos. Whatever order we have found in the universe, it is one that the mind has invented. Whatever attributes we have found in the universe, these are the ones that the mind has conceived. The universe is neither big nor small, neither beautiful nor ugly. The glow of mind fills the void and makes its space divine. This is what ecological consciousness is about. It is about developing and articulating the capacities that enable us to dwell in a reverential universe, and ultimately to live in grace.

It is only from the time when *some* minds tuned themselves into the sacramental or divine mode that the universe could be experienced as divine. When such superbly tuned minds appeared in India, they created *The Upanishads*. When such minds appeared in the ancient Hebrew world, they created the Bible. In ancient Greece these minds were exemplified by Pythagoras and Plato, who talked about the god within, our inner divinity.

These minds, who had originally projected their divinity upon the cosmos, were so delighted with their creation that they decided to attribute this divinity to the cosmos. They made the cosmos itself divine. They claimed that the divinity is in the cosmos, especially after they invented Brahma, Jehovah and godhead, whom they conceived as the Absolute Ground for Being from which everything springs.

What I am proposing, on the other hand, is the natural conception of divinity, or the *noetic* conception of divinity, as the mind is the creator of all orders, including the divine or spiritual order. Sacredness is an attribute of the mind, not an attribute of the cosmos. Only when we approach the universe with a reverential attitude and behold it with a mind that is sacred, do we find the universe sacred.

Whatever we have said about reverence also applies to spirituality. If we assume that the universe contains nothing but dumb matter and the only way to know this universe is through physical knowledge, we shall obviously not find any traces of spirituality in the universe because our assumptions and our language have ruled it out.

If we assume, on the other hand, that the universe is spiritually alive and that we are spiritual agents, and if we act on these assumptions, we shall find ample evidence that the universe is transphysical and transbiological or simply spiritual, as so many people of spiritual inclinations – and with the courage to assume that the universe is a spiritual place – have found throughout history.

OF EVOLUTION AND HOPE

Let us briefly discuss the last two characteristics, the evolutionary and participatory aspects of ecological consciousness. The beauty of our universe is inherently connected with the beauty of evolution, this awesome process within which amazingly creative forces have cooperated to bring about one miracle of life after another. The French philosopher Henri Bergson has captured the essence of this process by coining the phrase 'creative evolution'. His compatriot, Pierre Teilhard de Chardin was a rhapsodic bard singing countless hymns to creative evolution. Teilhard did much to convince us that evolution is not a dreary Darwinian process of slugging from one chance to another chance, from one incomprehensible necessity to another. (I am referring particularly to Jacques Monod's book, *Chance and Necessity*, which attempted to reinforce the materialist world-view and the Darwinian concept of evolution.) No, evolution is not a stupid and chancy process of stumbling upon one beneficial variation after another. Evolution is so exquisite in its mode of operation that it could be called divine. I, myself, have no difficulty in accepting the idea that God is evolution, and evolution is God – as I have argued in chapter 4.

The universe has brought forth life to celebrate itself. We are part of its glory. We neither deny our special place nor are unduly arrogant about it. It has just so happened that we are part of the flowering of the universe. To deny this special place to *Homo sapiens* in the name of the ideology of anti-anthropocentrism is a folly based on a new form of misanthropy; it is, in fact, a form of inverted human arrogance. The quintessence of evolution is that it brings about higher and higher forms of life. We are one of these higher forms.

In brief, ecological consciousness is evolutionary *par excellence*. Why is the acceptance of creative evolution important to the overall structure of ecological consciousness? And furthermore, why is the right comprehension of evolution important to our sense of the future and to our destiny?

For three reasons. Firstly, a sensitive reading of evolution informs us that the universe is quintessentially unfinished, that we are essentially unfinished. From this it clearly follows that we shall further unfold as the cosmos will unfold. Another important conclusion that follows is that we have somewhere to go, that we have a stupendous future in front of us. We are still toddlers in the cosmic playpen. We shall mature, we shall take our destiny more resolutely into our hands; we shall become less stupid, less vulgar, less consumptive, more frugal and more wise. This is what an intelligent reading of evolution informs us of in the first place.

Secondly, an intelligent reading of evolution informs us that evolution is a *divinizing* agent, that it transforms matter into spirit. Although some people claim that consciousness must have been present in matter and the entire universe from the beginning (otherwise how could it come to exist), I do not accept this view. Consciousness is *emergent*. It came to exist at a certain point in the development of matter. Even if I could agree that consciousness could have been there – deeply hidden in the inner layers of matter, waiting to be released – I would still argue that this process of *releasing* it from the bondage of matter was so extraordinary and so creative that we can easily talk about the creation of consciousness, not merely its unveiling. If some people wish to talk about the divinizing of matter as a process of unveiling what is potentially there, I have no problem with their language.

Thus at a certain juncture of evolution (perhaps with the first amoebas), dim consciousness was born. Then more articulated forms of consciousness appeared. Then self-consciousness appeared. Then spirituality and divinity appeared, as exquisite forms of self-consciousness reflecting on the possibilities of its own structure. Sacredness and spirituality – to reiterate the point – are an element of the structure of our consciousness, as it refines itself and transcends the physical and the biological. Evolution is thus a subtle process of divination, of transforming matter into spirit, transforming consciousness into self-consciousness, self-consciousness into sacred consciousness.

An intelligent reading of evolution is important for a third reason. It enables us to formulate what I would like to call the 'middle road', which lies between religious consciousness and technological or materialist consciousness. The former claims that all divinity and spirituality is God-given and represents a reflection of God's divinity. The latter claims that consciousness is a function of matter (Marxism) and that spirituality and divinity are delusions or fictions of the human mind. We claim that spirituality is an aspect of unfolding evolution. Let us notice (in passing) that Darwinism, which is an extension of the materialist world-view, does not have any answer as to how self-consciousness emerged and how to explain spirituality and the sacred.

In proposing the middle road, ecological consciousness is not unlike Buddhism, which does not evoke any notion of god but which nevertheless assumes that we are spiritual and divine creatures, and that through our own work, through our own karma, we can attain levels of high spiritual enlightenment. This enlightenment comes through concerted practice, through the right tuning of the mind, through evolving one's own psychic capacities. We are all possessors of our own divinity, but to release it from the bondage we need to perform a Herculean work of self-cleansing, and then of tuning ourselves to the most evolved forms of human consciousness. Somehow in our busy lives we tend to forget that all progress, and especially evolutionary progress, has been made by the painstaking development of consciousness – which holds the key to all our future.

In brief, creative evolution, as a component of ecological consciousness, is important for at least three reasons. It enables us to see that we are essentially unfinished. It enables us to make sense of the turbulent past while looking forward to the immense promise of the future. It enables us to follow the middle road of natural divinity within which we conceive of ourselves as both corporal and spiritual, both rational and mystical, all within the bounds of natural evolution.

Let us finally attend to the third characteristic of ecological consciousness, its participatory character, which spells out the participatory mind. The astrophysicist John Archibald Wheeler writes:

> The universe does not exist 'out there' independent of us. We are inescapably involved in bringing about that which appears to be happening. We are not only observers. We

are participators. In some strange sense this is a participatory universe.

The idea of the participatory universe would be void and meaningless if it did not create the participatory mind as its partner in co-creation. We never describe the cosmos as it is. We always partake in what we describe. Our mind invariably and tirelessly elicits (through the various faculties and sensitivities we possess) from the amorphous primordial data of the universe. A given condition is never given as such; it is always mediated, moulded, shaped and determined by the mind.

Ecological consciousness signifies a mind tuned into the biotic world in a holistic and symbiotic manner. But it also signifies a mind tuned into the whole universe in a co-creative manner. A true revolution in our views on the mind is not merely a denial that the mind is a blank surface in a mechanistic universe; or what comes to the same, a denial of the idea of the mind as a mirror photographing nature as it is. If objectivity no longer holds (and it does not), then an altogether new concept of mind and of consciousness must be evolved to make sense of the participatory universe in the state of co-creation. The idea of the participatory mind (as an aspect of ecological consciousness), is the element of the architecture of eco-philosophy that ties our more specific eco-concerns with more sublime reaches of the universe in the process of creation.

The participatory mind is the creator of all orders, including the spiritual order. We do not know whether there is any order in the universe except that which the mind finds there, that is to say, which the mind *imposes* on the universe. We know, however, that if we assume the universe to be mechanistic in nature we shall read it in a mechanistic way. On the other hand, if we assume that the universe is reverential in nature, then we can interact with it reverentially. How we treat the mind is exceedingly important, for on this depends how we treat the universe and ourselves. All changes in the universe start with changes in our mind.

Our mandala is now completed. We have articulated six essential characteristics of ecological consciousness. This consciousness is holistic, qualitative, spiritual, reverential, evolutionary and participatory. These characteristics form one coherent whole. They also partake in the making of eco-cosmology; on the one hand, they are consequences of this cosmology, and on the other hand, they are the constituents of this cosmology.

There is one element of ecological consciousness that is not in the mandala but which is of vital importance. This element is *hope*. Dante identified hope with heaven and hopelessness with hell. Indeed, the inscription at the Gate of Hell in *The Divine Comedy* reads 'Hope abandoned'.

Hope is crucial to life lived in beauty. Hope is so important to our conception of the holistic universe and to our attitude of reverence for life that it must be a part of ecological consciousness. When hope crumbles, everything crumbles.

Our fractured and atomized consciousness has created the spiritual vacuum, in the wake of which hopelessness creeps in naturally. This is a form of hell itself – to live in the world without hope. Technological consciousness is unable to provide any structure for hope. For this reason alone it condemns itself as an insufficient vehicle for living.

Hope is not wishful thinking but a vector of continuing transcendence. Hope is part of the skeleton of our existence. Hope is part of our ontological structure. Hope is the oxygen for our souls. Hope is a reassertion of our belief in the meaning of the universe. Hope is a precondition of all meaning, of all striving, of all action. To embrace hope is a form of wisdom. To abandon hope is a form of hell. In an essential sense, hope pervades the entire structure of ecological consciousness.

A CODA ON THE PRACTICAL EXERCISE LEADING TO THE ACQUISITION OF ECOLOGICAL CONSCIOUSNESS

The process of the greening of consciousness is going on, but a word of caution would be advisable. This process will not happen automatically. It will require great energy and will-power on our part. Nor will evolution proceed automatically and effortlessly to its happy end – the omega point – without our help. We are evolution conscious of itself – that is true. This realization imposes on us the enormous and beautiful burden of taking the responsibility for all there is, for our own future and for the future of evolution.

It is sometimes assumed that evolution is working with swiftness and necessity and will deliver us to some promised land no matter what. It is also assumed that evolution is making such stupendous breakthroughs that it is only a matter of years before the bulk of humanity becomes enlightened, and we shall all be on the right path. These views I consider altogether much too optimistic and indeed dangerously naïve.

To change the human consciousness presently stuck in the mechanistic mores will require a painstaking effort. It will require some kind of revolution. This is how Ionesco defines revolution: 'Revolution is a change in the state of consciousness.' Incidentally, revolutions which did not work, including the Soviet Revolution, are the ones which failed to create a new consciousness.

Now let us consider another subtle point. Why is a change of consciousness so difficult on the individual level? Because the organism sees it as a challenge to its identity. We are comfortable in our old niches – whatever they are. Old niches are tantamount to stability. New consciousness implies instability to begin with, even if it leads to liberation and new freedom in the long run.

Compassion, reverence and holistic thinking will not be given to us *deus ex machina* but will require diligent systematic work. I call these special exercises, which lead to the acquisition of ecological consciousness eco-yoga. There is no space to explain here the principle of eco-yoga, particularly as any yoga is to be practised rather than talked about.[2] Spiritual practices of old venerable traditions may be of considerable help. Many of the yogic systems presently available may be of considerable help too, if discriminatingly practised.

Working on ourselves is always a painful venture, even in spiritually orientated societies. It is particularly protracted and painful in indulgent societies, such as ours. Yet, if we are going to make it, we need to change within. We are still postponing this act of cleaning ourselves up, which will be part of the act of cleansing the environment at large. Subconsciously we are waiting for a Messiah who will do it for us, or for some wonderful technology that will miraculously solve our problems. This is the heritage of Messianic thinking and of technological thinking. The heritage of responsible thinking, on the other hand, tells us that we have to do it ourselves. And we shall because we are intelligent and resourceful beings – particularly when pressed against the wall. And we are pressed against the wall! We need to realize this fully, then look to our deeper resources and begin to reconstruct ourselves, like the phoenix from the ashes.

CONCLUSION

A change in the nature of consciousness signifies a change in the description of the world. We are here at such a point of change at present. Through the acquisition of ecological consciousness we are changing the nature of our perception, we are changing the nature of our knowledge, we are changing the nature of external reality. As we filter and sculpt reality differently, so we receive it differently.

We have argued that ecological consciousness presupposes and entails a cosmology which I call eco-cosmology (see chapter 1). The relationship between ecological consciousness and eco-cosmology is a very close one. One co-defines the other. One needs the other for its existence. It may be said that ecological consciousness represents the interiorization of the principles of eco-cosmology. Eco-cosmology, on the other hand, needs ecological consciousness as its articulator, as its alter ego, as its twin in the world of consciousness. Cosmology does not exist by itself; it comes to existence when it is articulated by the appropriate modes of consciousness. Those appropriate modes of consciousness we have to acquire individually, by working on ourselves. Only then will ecological consciousness become a social reality.

Thus the whole cycle is completed. We envisage a new world-view. We attempt to make it coherent. We also attempt to justify it rationally. Then we attempt to see what kind of architecture of the mind is presupposed by our new world-view. And then we develop a form of consciousness that is attuned to a given cosmos – refining it, articulating it, sustaining it and being sustained by it. This is what all cultures do. Traditional cultures go through this process intuitively, being only dimly aware of the intricacies of the process. In our culture, however, with the enormous knowledge of past cosmologies and past consciousnesses, we can engage in the process of creating a new consciousness quite consciously and deliberately.

Yet most people will acquire ecological consciousness through osmosis, through various cycles of praxis, as it was customary for traditional cultures. Ordinary people in the industrial countries, being continually assaulted by various forms of the fallout of 'progress', have already connected in their minds that industrial waste means pollution, pollution means poison. They have also become increasingly aware that recycling is good. Thus many elementary practices have left behind the sediment of values: pollution – bad; recycling – good; ruining the planet – disastrous. Those realizations will in time lead to larger realizations: frugality – good; conspicuous consumption – bad; healing the planet – our responsibility. At a later time still – after more subtle changes have occurred in the sphere of values – ecological lifestyles will be developed, and ecological consciousness will become a reality in action.

Action is important and always will be. But logos is equally important. For logos is a powerful form of praxis. Some of the urgent actions that we deemed so important ten years ago are now completely forgotten; and if remembered seem so trivial and insignificant. On the other hand, some deep insights or beautiful thoughts that we discovered years ago, may nourish us for years and years. We should not wish to diminish the sphere of action; but we should not wish to deify it either. Right action inevitably springs from deep reflection. Consciousness and action must be fused together. What cements them together are values, in our case ecological values. In a world-view that is well-ordered, values are intermediaries between consciousness and action. Values infuse meaning into action, on the one hand, and translate consciousness into modes of social praxis, on the other.

And let us be perfectly aware that we are not engaged in a crusade to change consciousness as if it were an abstract entity independent of the rest of our lives, and independent of the various forms of economic and social practice. Our consciousness works through an enormous number of the tributaries of life. In changing consciousness we are addressing ourselves to these various tributaries. So taxing polluters will help. Taxing motorists who drive more than an average number of miles will help. Recycling on a large scale will help. But finally we have to go to the source of our trouble – to ourselves. *We have to recycle our minds*. We have to acquire a new consciousness that is non-polluting, sane, sustainable and compassionate. Thus we are led back to the serene garden of ecological consciousness.

We are alone while being born and while dying. We are also alone in making epochal breakthroughs. We are united with the stupendous tapestry of life and evolution in so many ways. Yet when we struggle with our inner self, when we attend our soul and attempt to make it more infused with light, we are often alone. The acquisition of ecological consciousness must proceed on various levels of social praxis, but also within our own deeply individual lives.

Solitude is the mother of perfectibility.
Courage is the fire of the soul.
Hope is the everlasting spring of sustenance.

The right path is recognizing the wisdom
Of the cosmic constraints.
Ecological consciousness is finding a structure
Of sustainable and compassionate beauty
That encompasses it all.

Afterword

The late 1980s saw incredible political changes in Eastern Europe. And the term 'incredible' is just right, for who would have believed that communism would have collapsed so swiftly and, in most of the countries, bloodlessly? All these traumatic changes have simply shown that *the human agenda is forever open*.

A quarter of a century ago we would not have dreamt that the epoch of ecology was around the corner; now we can see that the ecological epoch is dawning on us. The age of Enlightenment was the epoch of transition from the age of religion to the age of science, and ultimately to the age of technology triumphant. We are now entering the Ecological Epoch which will be a prelude to a new age of the spiritual liberation of humankind.

My way to eco-philosophy led through many indirect paths, via analytical philosophy, philosophy of science, philosophy of technology and philosophy of man. I came to the United States in 1964, still convinced that America was the harbinger of the future, and that technological progress was the key to all progress. My first months in Los Angeles were intoxicating if somehow bewildering. I was constantly told that there is no further west, and that Los Angeles was *it*. Somehow, my life in Los Angeles did not quite appear to be the paradise I was told I was in. I also began to notice that this wonderful progress was not as wonderful from close range as it appeared from a distance. The freeways were always crowded. If a new one was built, it was clogged in a few months. I was told by a knowledgeable civil engineer that freeways do not relieve traffic, rather they *attract* traffic. This was quite a novel way of looking at things. It began to dawn on me that this may be the case with our wonderful technologies – they do not satisfy our needs but increase them.

Then came the hippie revolution while I was still living in Los Angeles. I actually lived off Sunset Boulevard where all manner of things were happening. Listening carefully to the voices of the angry young men at the time, I came to the conclusion that the problems of Western civilization were much deeper than we cared to acknowledge. The rebellious young souls were persistently asking me: 'You are a philosopher. Tell us where we have gone wrong.' The question was not *whether* we had gone wrong but *where* and *when*. I was disturbed by these questions as I had assumed, along with others, that we were the most rational and therefore most perfect of civilizations.

While searching for the causes of Western civilization backfiring upon itself, I scrutinized the last four centuries of Western culture and especially Western philosophy. I came to the conclusion that the fault was in the very blueprint, right there, in the seventeenth-century mechanistic philosophies which laid the foundations for the whole civilization. It was then that we formed the assumptions that the universe is a machine, that knowledge is power and that nature is ours for exploitation and plunder. We simply conceived of a wrong idiom for our interaction with nature.

This idiom led in time to the articulation of material progress as our new deity and to the elevation of the Faustian man as our new ideal. Driving relentlessly the chariot of material progress during the last two centuries, we have inadvertently destroyed nature and a good deal of the texture of society and of our individual lives.

I slowly came to the conclusion that progress is a loaded dice and that technology which pushes progress regardless of good or bad consequences is a malevolent entity; and that 'objective' science which allows itself to be used for the purposes of exploitative technology is not neutral but an accomplice of the whole dubious scheme. Thus I came to view the entire underpinning of Western civilization as unsound, and very dangerous. I saw the beauty and the potency of technology. But I also saw that technology was condemning itself by the fruit it was bearing:

desolate environments, atomized society, and individual alienation all being the consequences of a certain way of reading the world and interacting with it. These views on technology and progress, which were also extended to science and its underlying mechanistic cosmology, I formed in the late 1960s and early 1970s. They were all foreshadowed by my paper written and published in 1970, and entitled: 'Technology – The Myth Behind the Reality'; and then continued in such papers as 'The Scientific World-view and the Illusions of Progress', 'Science in Crisis', and 'Does Science Control People or Do People Control Science?'

While thinking about the vicissitudes of our mechanistic world-view I was struck many a time by how slow we are in learning. We consider ourselves clever, quick and intelligent people. Yet we learn awfully slowly from our past mistakes, and we are so reluctant to see and admit that the whole blueprint of our civilization is ridden with shortcomings, is now in tatters and has always been lamentably lacking in vision.

In the late 1970s the conservative backlash started to assert itself strongly. Many of my colleagues, who were once in the forefront of radical thinking, began to 'adjust'. It became 'unpopular' to uphold radical, ecology-orientated views. I saw no reason to 'adjust' my views as the whole matrix, which I recognized as defective and misfiring, was still intact. One 'knowledgeable' person at the time, upon learning that I was working on a philosophy in the ecological key, said to me: 'Why do you do that? Ecology is no longer in vogue.' This struck me as funny. How can ecology not be in vogue (I thought) while we were not solving any of the major problems caused by polluting technologies? So, I quietly went on developing my ideas. In 1981 I published the book *Eco-philosophy, Designing New Tactics for Living*, which was subsequently translated into twelve languages.

With the permission of the reader I will recall the circumstances leading to the publication of this book, which was probably the first systematic treatise on eco-philosophy. The first outline of my eco-philosophy was sketched in the following circumstances. On 20 June 1974, I was invited by the Architectural Association

School of Architecture in London (one of the best schools of architecture in Europe) to participate in the symposium entitled 'Beyond Alternative Technology'. We were already convinced, at this time, that the ecology movement had somehow burned itself out. Building windmills and insisting on soft technology was not enough. So four of us took the floor to ask ourselves, 'Where do we go from here?' Each of us had exactly ten minutes to deliver his message. What can you say in ten minutes? Not much. And yet you can say quite a lot. Instead of analysing the shortcomings of the ecology movement I decided to make a leap forward and ask myself, 'What is most troubling in the foundations of our knowledge, and what other foundation could we imagine as the basis for new thinking and a new harmony?' The sketch that I delivered was entitled *Ecological Humanism*. In it were formulated the major ideas that became the backbone of my eco-philosophy. It happens rarely that one is aware of the exact point of a new departure. It was perhaps a coincidence that the Architectural Association immediately published my text. Only recently did I find the original text and was rather amazed that all the seeds of eco-philosophy were contained in this presentation.

At the beginning of my address I said:

> Oswald Spengler has written that 'Technics are the tactics for living'. This is a very useful phrase indeed. I shall take advantage of it while stating our dilemma and while searching for possible solutions. Modern technology, – or better, Western technology – has failed us not because it has become economically counter-productive in the long run; and not because it has become ecologically devastating, but mainly because it has forgotten its basic function, namely that all technics are, in the last resort, tactics for living. Because modern technology has failed us as a set of tactics for living, it has also proved in the process to be economically counter-productive and ecologically ruinous.

At the end of my 10 minutes, I maintained:

Ecological humanism points towards social relationships based on the idea of sharing and stewardship rather than owning things and fighting continuous ruthless battles in open and camouflaged social wars. In short, ecological humanism is based on a new articulation of the world at large:

- it sees the *world* not as a place for pillage and plunder, an arena for gladiators, but as a *sanctuary* in which we temporarily dwell, and of which we must take the utmost care;
- it sees man not as an acquisitor and conquistador, but as a *guardian* and *steward*;
- it sees *knowledge* not as an instrument for the domination of nature, but ultimately as a technique for the refinement of the soul;
- it sees *values* not in pecuniary equivalents, but in intrinsic terms as a vehicle which contributes to a deeper understanding of people by people, and a deeper cohesion between people and the rest of creation;
- and it sees all these above mentioned elements as a part of the new tactics for living.

These ideas turned out to be fruitful, and during the next fifteen years I built a whole tree of new philosophy around them.

As the reader may be aware, providing *systems* of philosophy has been unfashionable in the twentieth century. However, I seem to have provided such a system in this book. My excuse is as follows. While trying to understand in some depth the major dilemmas of the twentieth century, I have been continually frustrated by the inadequacy and indeed glibness of answers given. So, out of desperation, I started to devise my own answers. After a while I realized that I could not shed a significant light on specific phenomena until I shed a new light on them all. Thus the system of eco-philosophy was born.

We are engaged in one of the crucial battles for the survival of the species. What we need are views that empower and affirm us, drive, the human species was bound to make mistakes – particularly when it became intoxicated with the power of new technologies and the power of reason itself. But we must not look at the human species from the perspective of its most destructive period. Moreover, during the last 150 years, when one segment of humanity was engaged in a systematic containment and destruction of nature, the rest of the human species coexisted with nature rather peacefully.

Let us take our clue from the most illustrious exemplars of humankind – Buddha, Jesus, Gandhi, Schweitzer, Mother Teresa, Martin Luther King and their ilk. And let us celebrate with those lights the glory of the human condition. The human agenda is forever open. Let us celebrate and articulate it in a way that is worthy of the magnificent species that we are. Let us welcome the dawn of the Ecological Epoch.

Notes to the text

CHAPTER 1

1 John D. Barrow, 'Life, the universe, and the anthropic principle', *The World and I*, August 1987, p. 182.

2 Willis Harman, 'Scientific positivism, the new dualism, and the perennial wisdom', *Scientific and Medical Network Newsletter*, Fall 1986, p. xx.

3 Nicholas Copernicus, *De Revolutionibus* (On the Revolution of the Spheres), Proemium 21.

4 See, for instance, G. Reichel-Dolmatoff, *Amazonian Cosmos*, Chicago University Press, 1971.

5 David Hume, *An Enquiry Concerning Human Understanding* (1748), ed. L. A. S. Bigge (1902), revised edition Oxford University Press, 1975.

6 Freeman Dyson, *Disturbing the Universe*, Harper & Row, 1981.

7 See John D. Barrow and Frank J. Tipler, *The Anthropic Cosmological Principle*, Oxford University Press, 1986.

8 Teilhard de Chardin, *The Phenomenon of Man*, Harper, 1959 (first published in French in 1957). For further discussion see Henryk Skolimowski, *Eco-philosophy*, Marion Boyars, 1981, and *The Theatre of the Mind*, Theosophical Publishing House, 1985. (By accepting the overall thrust of Teilhard's reconstruction of evolution, we do not necessarily accept every tenet Teilhard maintained about evolution.)

9 John Archibald Wheeler (see note 5, chapter 5).

10 For further discussion see Henryk Skolimowski, *The Theatre of the Mind*, Theosophical Publishing House, 1985, and 'The interactive mind in the participatory universe', in *The Real and the Imaginary*, ed. Jean Charon, Paragon House, 1987.

11 David Bohm, *Wholeness and the Implicate Order*, Routledge, Chapman & Hall, 1980.
12 For further discussion of 'reverence' see Henryk Skolimowski, *Eco-theology: Towards a Religion For Our Times*, Eco-philosophy Publications, 1985.
13 For further discussion of eco-ethics see Henryk Skolimowski, 'Eco-ethics as the foundation of conservation', *The Environmentalist*, 4, 1984, supplement no. 7, and 'Ecological values as the foundation for peace', *Ecospirit*, II, no. 3, 1986.

CHAPTER 2

1 From *Choruses From 'The Rock', Collected Poems, 1909–1962*, Faber and Faber Limited, 1974.
2 See Fritjof Capra, *The Tao of Physics*, Bantam, 1984.
3 E. F. Schumacher, *Small is Beautiful*, Harper & Row, 1975.
4 For further discussion see Henryk Skolimowski, 'Rationality, economics and culture', *The Ecologist*, June 1980.
5 Katie Kelly, *Garbage: The History and Future of Garbage in America*, Saturday Review Press, 1973, p. 41.
6 See especially Ivan Illich, *Energy and Equity*, Harper & Row, 1974. The Gulf War has proved this point only too eloquently.
7 A. I. Solzhenitsyn, The Harvard Commencement Address, 1978.
8 See Teilhard de Chardin, *The Phenomenon of Man*, Harper, 1959 (first published in French in 1957).
9 Philip Toynbee, The *Observer*.

CHAPTER 3

1 From *Pensées*, no. 23.

CHAPTER 4

1 Immanuel Kant, *The Groundwork of the Metaphysics of Morals*, II 67, tr. H. J. Paton, Harper & Row.
2 See especially Karl Popper, *Conjectures and Refutations*. (It

should be mentioned at this point that although piercing in his critique of logical positivism, Popper was a positivist, in the broad sense of the term, himself; he sought solutions to all our dilemmas in cognition; his intellectual universe seems to be limited to cognition alone – a typically positivist limitation.)

3 What I have offered here is an ecological (or eco-philosophical) critique of Marxism, probably the first of its kind. A further analysis of Marxism along these lines is carried in my articles.

4 Albert Schweitzer, *Out of My Life and Thought*, NAL, 1953 (first published 1948).

5 Theodosius Dobzhansky, 'Advancement and obsolescence in science', *Great Ideas Today: A Symposium on Tradition*, Encyclopaedia Britannica Education, 1974.

6 John Archibald Wheeler, 'The universe as home for man', *American Scientist*, November–December 1974, p. 688 onwards.

7 R. H. Dicke, *Nature*, 192, 1961, pp. 440–41.

8 John Archibald Wheeler, op. cit., p. 689.

CHAPTER 5

1 See especially the writings of Carl Jung, Mircea Eliade, Joseph Campbell.

2 See especially Karl Marx's *The Philosophical and Economic Manuscripts of 1844*; when one reflects on the manuscripts one is struck by their naïvety.

3 Jean-Paul Sartre, 'Where I got it wrong on despair', a conversation with Benny Levy, the *Observer*, London, 20 April 1980.

4 Jean-Paul Sartre, op. cit.

5 See John Archibald Wheeler, 'The universe as home for man', *American Scientist*, November–December 1974.

6 Although Gregory Bateson hints at epistemology of life and talks about it in his *Mind and Nature: A Necessary Unity* (E. P. Dutton, 1979), he does not provide one. However, some of his ideas are constructive.

7 After having found myself too constrained and suffocated in university auditoria and rooms, which feel like cages rather than halls of learning, I searched and found a place in the mountains, in the village of Theologos on the island of Thassos in Greece, where I attempt to develop philosophy as homage paid to life. During the summertime, groups of people come to the village to take workshops with me on eco-philosophy and eco-yoga.

8 *The Upanishads*, tr. J. Mascaro, Penguin Books, 1970.

CHAPTER 7

1 John G. Neihardt, *Black Elk Speaks*, University of Nebraska Press, 1988.

2 For further discussion see Henryk Skolimowski, 'Forests as sanctuaries', *Dancing Shiva in the Ecological Age*, 1991.

3 Robert Pirsig, *Zen and the Art of Motorcycle Maintenance*, William Morrow & Co., 1974.

4 Hundertwasser, *The Vienna Manifesto*, exhibition catalogue, 1974.

CHAPTER 8

1 Lynn White, 'The historical roots of our environmental crisis', *Science*, 155, no. 2767, 1967.

2 Reverence for life is very much Schweitzer's concept. Yet Schweitzer's notion of reverence was rather too anthropocentric and too bound to traditional Christian ethics. In our usage 'reverence' has a more universal scope, as it includes all beings.

3 For further discussion of intrinsic values, see Henryk Skolimowski, 'In defense of eco-philosophy and of intrinsic value: A call for conceptual clarity', *The Trumpeter*, 3, no. 4, Fall 1986.

4 Aldo Leopold, *A Sand County Almanac*, Ballantine, 1986.

5 For further discussion of responsibility, see Henryk

Skolimowski, *The Theatre of the Mind*, Theosophical Publishing House, 1985, chapter 19.

6 See Henryk Skolimowski, 'Eco-ethics as the foundation of conservation', *The Environmentalist*, 4, supplement no. 7, 1984.

CHAPTER 9

1 For further discussion of reverential thinking, see Henryk Skolimowski, *The Theatre of the Mind*, Theosophical Publishing House, 1985, chapter 10.

2 I have conducted workshops on eco-yoga during the summer months in a mountainous village on the island of Thassos, in northern Greece. (See also note 7, chapter 5.)

ARKANA – NEW-AGE BOOKS FOR MIND, BODY AND SPIRIT

A selection of titles

ARKANA – NEW-AGE BOOKS FOR MIND, BODY AND SPIRIT

A selection of titles

The Revised Waite's Compendium of Natal Astrology
Alan Candlish

This completely revised edition retains the basic structure of Waite's classic work while making major improvements to accuracy and readability.

Aromatherapy for Everyone Robert Tisserand

The therapeutic value of essential oils was recognized as far back as Ancient Egyptian times. Today there is an upsurge in the use of these fragrant and medicinal oils to soothe and heal both mind and body. Here is a comprehensive guide to every aspect of aromatherapy by the man whose name is synonymous with its practice and teaching.

Tao Te Ching The Richard Wilhelm Edition

Encompassing philosophical speculation and mystical reflection, the *Tao Te Ching* has been translated more often than any other book except the Bible, and more analysed than any other Chinese classic. Richard Wilhelm's acclaimed 1910 translation is here made available in English.

The Book of the Dead E. A. Wallis Budge

Intended to give the deceased immortality, the Ancient Egyptian *Book of the Dead* was a vital piece of 'luggage' on the soul's journey to the other world, providing for every need: victory over enemies, the procurement of friendship and – ultimately – entry into the kingdom of Osiris.

Yoga: Immortality and Freedom Mircea Eliade

Eliade's excellent volume explores the tradition of yoga with exceptional directness and detail.

'One of the most important and exhaustive single-volume studies of the major ascetic techniques of India and their history yet to appear in English' – *San Francisco Chronicle*

ARKANA – NEW-AGE BOOKS FOR MIND, BODY AND SPIRIT

A selection of titles

Weavers of Wisdom: Women Mystics of the Twentieth Century Anne Bancroft

Throughout history women have sought answers to eternal questions about existence and beyond – yet most gurus, philosophers and religious leaders have been men. Through exploring the teachings of fifteen women mystics – each with her own approach to what she calls 'the truth that goes beyond the ordinary' – Anne Bancroft gives a rare, cohesive and fascinating insight into the diversity of female approaches to mysticism.

Dynamics of the Unconscious: Seminars in Psychological Astrology Volume II Liz Greene and Howard Sasportas

The authors of *The Development of the Personality* team up again to show how the dynamics of depth psychology interact with your birth chart. They shed new light on the psychology and astrology of aggression and depression – the darker elements of the adult personality that we must confront if we are to grow to find the wisdom within.

The Myth of Eternal Return: Cosmos and History Mircea Eliade

'A luminous, profound, and extremely stimulating work . . . Eliade's thesis is that ancient man envisaged events not as constituting a linear, progressive history, but simply as so many creative repetitions of primordial archetypes . . . This is an essay which everyone interested in the history of religion and in the mentality of ancient man will have to read. It is difficult to speak too highly of it' – Theodore H. Gaster in *Review of Religion*

The Second Krishnamurti Reader Edited by Mary Lutyens

In this reader bringing together two of Krishnamurti's most popular works, *The Only Revolution* and *The Urgency of Change*, the spiritual teacher who rebelled against religion points to a new order arising when we have ceased to be envious and vicious. Krishnamurti says, simply: 'When you are not, love is.' 'Seeing,' he declares, 'is the greatest of all skills.' In these pages, gently, he helps us to open our hearts and eyes.

ARKANA – NEW-AGE BOOKS FOR MIND, BODY AND SPIRIT

A selection of titles

A Course in Miracles: The Course, Workbook for Students and Manual for Teachers

Hailed as 'one of the most remarkable systems of spiritual truth available today', *A Course in Miracles* is a self-study course designed to shift our perceptions, heal our minds and change our behaviour, teaching us to experience miracles – 'natural expressions of love' – rather than problems generated by fear in our lives.

Sorcerers Jacob Needleman

'An extraordinarily absorbing tale' – John Cleese.

'A fascinating story that merges the pains of growing up with the intrigue of magic . . . constantly engrossing' – *San Francisco Chronicle*

Arthur and the Sovereignty of Britain: Goddess and Tradition in the Mabinogion Caitlín Matthews

Rich in legend and the primitive magic of the Celtic Otherworld, the stories of the *Mabinogion* heralded the first flowering of European literature and became the source of Arthurian legend. Caitlín Matthews illuminates these stories, shedding light on Sovereignty, the Goddess of the Land and the spiritual principle of the Feminine.

Shamanism: Archaic Techniques of Ecstasy Mircea Eliade

Throughout Siberia and Central Asia, religious life traditionally centres around the figure of the shaman: magician and medicine man, healer and miracle-doer, priest and poet.

'Has become the standard work on the subject and justifies its claim to be the first book to study the phenomenon over a wide field and in a properly religious context' – *The Times Literary Supplement*

ARKANA – NEW-AGE BOOKS FOR MIND, BODY AND SPIRIT

A selection of titles

Being Intimate: A Guide to Successful Relationships
John Amodeo and Kris Wentworth

This invaluable guide aims to enrich one of the most important – yet often problematic – aspects of our lives: intimate relationships and friendships.

'A clear and practical guide to the realization and communication of authentic feelings, and thus an excellent pathway towards lasting intimacy and love' – George Leonard

Real Philosophy: An Anthology of the Universal Search for Meaning Jacob Needleman

It is only in addressing the huge, fundamental questions such as 'Who am I?' and 'Why death?' that humankind finds itself capable of withstanding the worst and abiding in the best. The selections in this book are a survey of that universal quest for understanding and are particularly relevant to the awakening taking place in the world today as old orders crumble. The authors call it Real Philosophy.

The Act of Creation Arthur Koestler

This second book in Koestler's classic trio of works on the human mind (which opened with *The Sleepwalkers* and concludes with *The Ghost in the Machine*) advances the theory that all creative activities – the conscious and unconscious processes underlying artistic originality, scientific discovery and comic inspiration – share a basic pattern, which Koestler expounds and explores with all his usual clarity and brilliance.

Whole in One: The Near-Death Experience and the Ethic of Interconnectedness David Lorimer

This prodigious study of the interconnectedness of creation draws on world-wide traditions of the after-life and detailed near-death experiences to posit an intrinsic moral order at work in the universe. 'This great book will take its permanent place in the literature of the spiritual renaissance in our time' – Sir George Trevelyan

ARKANA – NEW-AGE BOOKS FOR MIND, BODY AND SPIRIT

A selection of titles

A History of Magic Richard Cavendish

'Richard Cavendish can claim to have discovered the very spirit of magic' – *The Times Literary Supplement*. Magic has long enjoyed spiritual and cultural affiliations – Christ was regarded by many as a magician, and Mozart dabbled – as well as its share of darkness. Richard Cavendish traces this underground stream running through Western civilization.

One Arrow, One Life: Zen, Archery and Daily Life
Kenneth Kushner

When he first read Eugen Herrigel's classic *Zen in the Art of Archery* at college, Kenneth Kushner dismissed it as 'vague mysticism'; ten years later, he followed in Herrigel's footsteps along the 'Way of the Bow'. *One Arrow, One Life* provides a frank description of his training; while his struggles to overcome pain and develop spiritually, and the *koans* (or riddles) of his masters, illustrate vividly the central concepts of Zen.

City Shadows Arnold Mindell

'The shadow destroys cultures if it is not valued and its meaning not understood.' The city shadows are the repressed and unrealized aspects of us all, lived openly by the 'mentally ill'. In this compassionate book Arnold Mindell, founder of process-oriented psychology, presents the professionals of the crisis-ridden mental health industry with a new and exciting challenge.

In Search of the Miraculous: Fragments of an Unknown Teaching P. D. Ouspensky

Ouspensky's renowned, vivid and characteristically honest account of his work with Gurdjieff from 1915–18.

'Undoubtedly a *tour de force*. To put entirely new and very complex cosmology and psychology into fewer than 400 pages, and to do this with a simplicity and vividness that makes the book accessible to any educated reader, is in itself something of an achievement' – *The Times Literary Supplement*